Library of
Davidson College

COUNTER-MOVEMENTS IN THE SCIENCES

SOCIOLOGY OF THE SCIENCES

A YEARBOOK

Editorial Board:

G. Böhme, *Technische Hochschule, Darmstadt*
N. Elias, *University of Leicester*
Y. Elkana, *The Van Leer Jerusalem Foundation, Jerusalem*
R. Krohn, *McGill University, Montreal*
G. Lemaine, *Ecole des Hautes Etudes en Sciences Sociales, Paris*
W. Lepenies, *Free University of Berlin*
H. Martins, *University of Oxford*
E. Mendelsohn, *Harvard University*
H. Nowotny, *European Centre for Social Welfare Training and Research, Vienna*
H. Rose, *University of Bradford*
P. Weingart, *University of Bielefeld*
R. D. Whitley, *Manchester Business School, University of Manchester*

Managing Editor: R. D. Whitley

VOLUME III – 1979

COUNTER-MOVEMENTS IN THE SCIENCES

The Sociology of the Alternatives to Big Science

Edited by

HELGA NOWOTNY *and* HILARY ROSE

D. REIDEL PUBLISHING COMPANY

DORDRECHT : HOLLAND / BOSTON : U.S.A.
LONDON : ENGLAND

Library of Congress Cataloging in Publication Data

Main entry under title:

Counter-movements in the sciences.

    (Sociology of the sciences ; v. 3, 1979)
    Includes bibliographical references and index.
    1. Science—Social aspects—Addresses, essays, lectures.
I. Nowotny, Helga. II. Rose, Hilary, 1935– III. Series.
Q175.55.C68   301.5   79–12683
ISBN 90–277–0971–8
ISBN 90–277–0972–6 pbk.

Published by D. Reidel Publishing Company,
P.O. Box 17, Dordrecht, Holland

Sold and distributed in the U.S.A., Canada, and Mexico
by D. Reidel Publishing Company, Inc.
Lincoln Building, 160 Old Derby Street, Hingham,
Mass. 02043, U.S.A.

All Rights Reserved
Copyright © 1979 by D. Reidel Publishing Company, Dordrecht, Holland
and copyright holders as specified on appropriate pages
No part of the material protected by this copyright notice may be reproduced or
utilized in any form or by any means, electronic or mechanical,
including photocopying, recording or by any informational storage and
retrieval system, without written permission from the copyright owner

Printed in The Netherlands

TABLE OF CONTENTS

Introduction	vii
Biography of Contributors	xiii
HELGA NOWOTNY – Science and its Critics: Reflections on Anti-Science	1
JEROME RAVETZ – Anti-Establishment Science in Some British Journals	27
ROBERT FRANCK – Knowledge and Opinions	39
AGNES HELLER – Can the Unity of Sciences be Considered as the the Norm of Sciences?	57
INGO GRABNER and WOLFGANG REITER – Guardians at the Frontiers of Science	67
GERNOT BÖHME – Alternatives in Science – Alternatives to Science	105
OTTO ULLRICH – Counter-movements and the Sciences: Theses Supporting Counter-movements to the 'Scientisation of the World'	127
MICHAEL GRUPP – Science and Ignorance	147
MANFRED E. A. SCHMUTZER – It May Be That On Earth No-one Speaks the Truth	161
GEOFFREY PEARSON – Resistance to the Machine	185
T. J. PINCH and H. M. COLLINS – Is Anti-Science not-Science? The Case of Parapsychology	221
SUZANNE PETERS – Organic Farmers Celebrate Organic Research: A Sociology of Popular Science	251
HILARY ROSE – Hyper-reflexivity: A New Danger for the Counter-movements	277
Index	291

INTRODUCTION

Heretical thoughts in an orthodox series on sociology of the sciences? Devils and science between the covers of one book? Games with ambivalence to mask collective uncertainty? We anticipate similar future reactions from readers or reviewers when assessing the way in which this volume has been assembled. But writings on counter-science, like the history of colonialism, are usually written by the winners, therefore unequivocally partial and only too often lacking in social imagination. In seeking to redress the balance, we admit to having been fully receptive to the latter, of having displayed an unmeasured degree of sympathy with heretics and outsiders, including practising scientists, and to letting science defend itself. The antithetical relationship implied in the volume's title — Counter-movements in the Sciences — stands for what we regard as an ongoing, open-ended process. In collecting material for this volume, we have brought together voices speaking from different quarters:

— there are those who, although modestly claiming to speak only for themselves, have set out to question sacred assumptions of scientific faith or to cast doubt on well-known claims scientific knowledge holds over other forms of knowledge;

— others have undertaken to demonstrate the fragility, if not untenability of attempts at demarcation between science and other systems of belief or practice or shown that demarcations between different forms of rationality rest on other than methodological grounds;

— finally, those who wish to re-arrange, by mapping out some meta-point of surveillance, familiar territory, showing the need for rearrangement and assessing the possibilities for alternatives.

In a way, these are highly personalised accounts of a theme which has received little collective treatment by way of scholarly attention.[*] While none of the

---

[*] See, however, the collection of papers edited by G. Holton and W. Blanpied, *Science and its Public: The Changing Relationship*, D. Reidel, Dordrecht, Holland, 1976.

authors seem ready to pronounce themselves authoritatively on the outcome of an ongoing process in which science is confronted with seemingly 'irrational' belief systems, practices or criticism, a multi-faceted topic has been opened for intellectual discussion which we hope will be continued outside these pages, drawing in our readers as well.

Faced with the staggering phenomenological diversity of counter-movements and a terminology often coined with deprecatory intent, Helga Nowotny has set out in her introductory essay to bring some order into this unruly material. By pointing to the social rationality of counter-movements, areas of total or partial disagreement between science and counter-science can at least be roughly distinguished; the dual character of an antagonistic relationship becoming thereby apparent. The romantic tradition of anti-science, the secret aspirations of many pseudo-sciences and the critical voices inside science, speaking from a minority but strategic position, constitute the major forms treated in this essay.

Following Helga Nowotny's introductory essay is a chapter contributed by Jerome Ravetz, who brings to an analysis of some of the magazines of the radical and alternative science movement in Britain a sympathetic understanding of the scale and the nature of their critique of science. Sensitised by his own work on the social problems posed by scientific knowledge, Ravetz argues that these critiques exist substantially in isolation from the church of science. They exist as alternatives rather than as recognized heresies, partly because the new religion is less monolithic than the old (a point Heller also makes) which permits a certain diversity, and partly because their heresy is so radical that the subject of the critique, mainstream science, is unable to perceive it as such.

Whilst Ravetz, despite his evident sympathy towards these alternative visions, writes *about* them, the third chapter by Robert Franck writes firmly from *within* a search for alternative knowledges. Franck is concerned with what he sees as the exclusive claims of a scientific knowledge, which seeks to denigrate all other knowledges as mere opinion. He argues that this apparently clear cut division is not so well founded as it claims — a point made in other chapters — and instead serves to mask the relationship between the dominant class and this dominating knowledge system. Where the old ideology of the science of the 17th and 18th centuries demanded that only the Laws of Nature and Reason should be served, the new schism between subjectivity and objectivity technicalises knowledge until it becomes a machine within the productive system.

Speaking with the clear moral voice we have learnt to associate with the Budapest School, Agnes Heller returns, in the following essay, to the same problems which preoccupied Robert Franck. She reminds us of the two distinct programmes which have sought to bring about a unity of the sciences, and argues that the realisation of either of these would bring about a 'negative Utopia'. The former programme, under the banner of positivism, that aggressive attempt mainly launched by the philosophers of science, rather than scientists themselves, seeks to make all knowledges adopt the pattern of a mathematicised natural science; the latter turns to philosophy for unification. Whereas in pre-bourgeois society there was a profoundly unified value system, so that within each knowledge both differentiation and integration was expressed, with the advent of bourgeois society this sense of community is dissolved. Instead, Nature becomes a mere object to be manipulated, a new concept of objectivity is given birth and with this the language of natural science. She defends natural science, arguing that it is the extension of this epistemé over all others which is the hazard, not the science itself.

Because in her view there is a plurality of need, even in a future society, there should also be a plurality of epistemé. In contradistinction to some other contributors, Heller, whilst wishing to oppose the domination of the social sciences by the natural, does not seek to replace this knowledge by the domination of the social. Nonetheless, she sees an acute and pressing problem with the natural sciences, which she characterises as a degeneration of their original rationality, a problem however which is historical and not universal and which can therefore be reversed.

Lastly, she does not permit scientists merely to wait until the necessary social transformation has occurred and redefined the rationality of science, but instead reminds them of a personal moral duty to follow the Kantian injunction of refusing to carry out that research whose results require that one human being becomes the mere means of another.

The following essay by Ingo Grabner and Wolfgang Reiter surveys the ideological significance of the 'Frontiers of Science', so dear to the hearts of physicists and their followers. The authors argue that questions typical of frontier science make the validity of demarcation between it and the pseudo-sciences very doubtful. Their analysis, witty, competent and cautious, takes us through the cognitive, socio-economic and military geography of 'frontier country'. We meet the Devil with its many faces, are made to look into the magic mirror, observe the breeding of dinosaurs and are introduced to the clever machinations of monks and mandarins. Whether we return from this fascinating journey believing that the end of the myth of science is near, will depend on our relations with the Devil.

Alternatives in science and to science are the two major issues to which Gernot Böhme addresses himself. The question whether alternatives in science are possible is not likely to meet with neutral reception. This is at least the experience made in reflecting retrospectively on the project 'Alternatives in Science' carried out by the Starnberg group. Discussing the possibilities of such alternatives within a philosophical and epistemological frame takes us finally to the juxtaposition between the idea of useful science and 'good science', which according to Böhme must be realised on the level of cognition, if it can be brought about at all.

The essay by Otto Ullrich is a bold attempt to formulate in a programmatic statement the 'progressive' potential of counter-movements; that is, their ability to stem the tide of what Ullrich sees as a growing tendency of the 'scientisation' of the world. While some of his theses echo familiar themes from the ecological movement, others point in the direction of calling for more overt political action. It may not be totally coincidental that the two contributions which end in a pessimistic tone by highlighting the dangers residing in a possible alliance between science and an increasingly centralised and bureaucratised political power come from countries in which the rise of fascism had not only racist, but also technocratic roots, as Albert Speer's memoirs so vividly testify. While not repeating itself, history does not limit its lessons to singular occurrences either.

Next, we shift back from a philosophical perspective to that of a practising physicist. For Michael Grupp, science, like many old men, although ailing, is still at the peak of power. Its past successes have been achieved by working on what he calls 'simple' problems, but it is not so well equipped at all to deal with a new type of question, the large scale 'complex' problem with which we will be faced at an increasing rate in the future. Ignorance, far from being cause for shame, becomes a virtue, the only remedy against the danger of running head-on into global disaster. But are scientists ready to profess their ignorance and to renounce their false High Priests who pretend to know?

Introduction   xi

The following contribution is written in an apparently less serious vein. Manfred E. A. Schmutzer invites us to play supposedly innocent little games with him, as an outsider to the big, serious game of science. He takes us through various interpretations of scientific knowledge and seeks to persuade us to follow his relativistic message that we tend to find only what we are looking for. The historical development of science is recast in the light of this tenet, but suddenly we are made to realise that we are treading on dangerously realistic ground. Science is intrinsically intermingled with power and is an ally of the ruling class. The paper ends with a sinister warning of a future dominated by science as the omnipotent manager. The innocent little game has ended in a potential nightmare.

Rebellion against an earlier phase of scientific and technical rationality is celebrated in Geoffrey Pearson's essay discussing the historical significance of the Luddites. He sees them, as in E. P. Thompson's classical account of *The Making of the English Working Class* as less expressive of the first outburst of class anger of the new urban proletariat than the pain and rage of those whose traditions, based on a domestic economy, were being swept away by the new machines. But where Pearson seeks to follow Thompson in writing a history from below, so that there is a profound sympathy with the social rationality of the machine wreckers, in the last analysis Thompson endorses the organized labour movement as the necessary and superior response to capital. It is at this point that Pearson's analysis differs. This difference is most clear in the closing pages of his work where he considers the present threat of holocaust and pollution posed by the contemporary machine. He seems to invite us to draw a parallel between the social rationality of the machine wreckers and that of the new 'vandals'. Pearson thus places his hopes on spontaneous resistance, looking to those whom the authorities call 'vandals' to invent a new technology of destruction to weed out those machines which threaten human existence itself.

Trevor Pinch and Harry Collins turn to the task of documenting 'anti-science' as against 'science'. Essentially a paper written from within a yet incomplete research programme, its interests lie in the detailed — and not unamusing — evidence. They point to the ways in which the particular form of anti-science — pseudo-science, to use Nowotny's term — patterns itself on mainstream and successful science. They suggest that it is precisely because of the close correspondence between science and anti-science that parapsychology receives such savage treatment as a recognisable and identifiable heresy at the hands of orthodox science. Like Ravetz, they are aware of other strands within anti-science whose objectives are so different, that science and anti-science proceed in parallel tracks with no point of intersection or debate.

As a further example of the development of a new science, Suzanne Peters' contribution analyses the history of organic farming. She describes the founders of organic farming, Rudolf Steiner as a Christian mystic, and Sir Albert Howard as a no-nonsense agricultural expert, both of whom nonetheless shared a common antipathy to conventional agricultural science and openness to alternative approaches. From these origins she traces the development of 'organic' farming with its strong rhetoric of being a new science. Analysing the claims of this new 'science', she argues that, despite some gains from the new approach, they are premature. Instead, she notes the endurance of the claims and points, in ways echoing Pinch and Collins, to the skill and energy of the organic enthusiasts in articulating "relatively credible versions of the scientific ethos", which sustain their personal and collective faith in the new science.

The concluding paper by Hilary Rose strikes a more pessimistic note. Having welcomed what she characterises as the phases of the critique and auto-critique of the oppositional movements in science, she continues by pointing to the new dangers in the present phase of hyper-reflexivity. She argues that the intensely subjectivist pre-occupation with the conditions of the production of knowledge opens the door to an extreme version of sociological relativism, where notions of the truth and falsity of ideas are dissolved in an equality of discourse. Whilst not wishing to abandon the possibility of a new science, she seeks to return theory to being both based and tested on practice. She suggests that the new subjectivist radicalism expressed within the counter-movements and often reflected within the volume, may have its own sociological explanation, as a manifestation of the pain of the intellectuals increasingly trapped within the contradictions of an incorporated science.

These, then, are the contrary voices which, as the third *Yearbook of the Sociology of the Sciences*, speak of the new counter-movements.

HELGA NOWOTNY and HILARY ROSE

BIOGRAPHY OF CONTRIBUTORS

GERNOT BÖHME is professor of philosophy and philosophy of science at the Technische Universität Darmstadt. He has studied mathematics, physics and philosophy. He spent the years 1969–1977 at the Max Planck Institut zur Erforschung der Lebensbedingungen der wissenschaftlich-technischen Welt (studies of the life conditions of the scientific-technical world) in Starnberg where the project was carried out which he partly reviews for this volume.

ROBERT FRANCK is a philosopher particularly concerned with the problems of science and teaches at the Catholic University of Louvain.

INGO GRABNER and WOLFGANG REITER are two practising physicists who live in Vienna where they work in their respective institutes the 'Institut für Theoretische Chemie und Strahlenchemie der Universität Wien' and the 'Radiuminstitut'. They pursue the study of social problems of science and seek to develop their own alternative style which may include, as their acknowledgements reveal, lengthy discussions in congenial surroundings.

MICHAEL GRUPP has studied physics in Heidelberg and Grenoble. He participated as an expert against nuclear power in the official information campaign organized by the Austrian Government and, together with friends, has set up an 'Institut de Recherche Alternative' in Lodève in the south of France devoted to alternative research on energy problems.

AGNES HELLER is a leading member of the Budapest School. Her work, including *The Theory of Need in Marx* and *Renaissance Man*, has become increasingly known in Western Europe over the past decade. She is presently teaching at the University of La Trobe in Australia.

HELGA NOWOTNY has been preoccupied for a number of years with cognitive and social aspects in the study of science apart from her work as Director of the

European Centre for Social Welfare in Vienna. A sociologist with a Ph.D. from Columbia University, New York, she has lectured at the Institute for Advanced Studies in Vienna, the Wirtschaftsuniversität, Wien, the University of Graz and the University of Bielefeld.

In the process of editing this volume she has greatly enjoyed numerous stimulating conversations with several of the contributors and a feministic partnership with her co-editor, Hilary Rose.

SUZANNE PETERS is a sociologist from McGill University in Montreal who wrote her doctoral thesis on the alternative farming movement.

GEOFFREY PEARSON originally studied natural sciences at Cambridge. Subsequently, his interest turned towards psychiatric social work and the study of deviancy and is the author of the *Deviant Imagination*. He is actively involved in the National Deviancy Conference. He is presently a Senior Lecturer in the School of Applied Social Studies, University of Bradford.

TREVOR PINCH and HARRY COLLINS are both sociologists of science, teaching and researching at the University of Bath. Their paper contributed to this collection stems from their joint research. Harry Collins is also the convenor for the British Sociological Association's study group in sociology of science.

JERRY RAVETZ has long written in the history of science, but has always integrated his scholarly interests with a moral and political concern as the title of his book *Science and Its Social Problems* indicates. He has worked with a variety of groups active on these issues from the establishment Council on Science and Society, the radical British Society for Social Responsibility in Science, to the alternative groups he discusses in this paper.

HILARY ROSE has been writing for some years, usually with Steven Rose, within the sociology of science. Their first book was a critique of science policy and their most recent edited books explored the question of ideology in science. She has also been active within the radical science movement, and finding the women and science collective (of which Ravetz writes) the most congenial setting. She is a Professor in the School of Applied Social Sciences at the University of Bradford where she teaches social policy. Thus, in common with her co-editor Helga Nowotny her interests span from social welfare to science.

MANFRED E. A. SCHMUTZER studied engineering at the Technische Hochschule in Vienna and later turned to the social sciences. He studied and lectured in political science and sociology at the University of Essex and in Vienna. His latest 'little games' took him to the field of cognitive sociology. Together with others in Vienna, he is currently engaged in a project dealing with the decline of the belief in progress.

OTTO ULLRICH received a manual-technical training as communication mechanic and electrical engineer and has worked for industry before studying sociology, psychology and economics in Berlin and London. He holds a diploma in sociology and a Ph.D. in political science and is the author of *Technik und Herrschaft*. He lectures at the Freie Universität, Berlin. The main focus of his present work are the life conditions of the scientific-technical world.

SCIENCE AND ITS CRITICS:
REFLECTIONS ON ANTI-SCIENCE

HELGA NOWOTNY
Universität Bielefeld

1. In Defence of Counter-movements

When setting out to plan this volume, we had a reasonably clear idea of what we understood by counter-movements: social forms of protest, contemporary as well as historical, which were critical of scientific rationality and at times hostile to specific technological developments. We were interested in their salient characteristics, in terms of social origins and intellectual concerns, and wanted to analyze the specific forms which ideological clashes between scientific and other forms of knowledge could take. We expected rebellion and resistance against the dominant faith of scientific rationality and wanted to extend our intellectual concerns to an analysis of the kinds of rationalities which were being developed within or against the existing sciences.

What I was discovering, however, was that these counter-movements had already been subject to a certain amount of ridicule, contempt and hostility on the part of scientists who, largely in good faith, felt called upon to defend science against what they interpreted as a rising flood of irrationality. Allusions to, as well as explicit writings on, anti-science, pseudo-science and non-science, to cite only three favourite labels bestowed upon the counter-movements, were intended to contain them in advance and to ostracize or domesticate any attempts to deviate from the dominant form of scientific rationality. As the labels were used indiscriminately for opponents outside science, but occasionally also for colleagues, the scene to be surveyed presented itself as terminologically confusing, analytically messy and sociologically fertile, but difficult. At times it seemed to me that scientists engaged in their attacks on anti-science were obsessed with an enemy which made them see the devil on the wall behind every innocent concern, however legitimate.

Their relations to a largely lay public were beset by a lack of understanding and unwillingness for any form of empathy. Rationality stood against irrationality, facts against emotions.

But, however fascinating such a response, I could not deal exclusively with a reaction to a phenomenon without investigating the phenomenon itself. Ideally, this would have necessitated a research project in its own right. From my own limited experience (0) I had become convinced that forms of protest and attacks against science and technology had more to do with who is to control a future world and divergent views as to what it should look like, than a turning against science as such. In other words, I saw primarily a social and political conflict as the basis for such movements, which scientists — to whom such words may sound more alien — perhaps deliberately interpreted as epistemological conflict or, in reductionist terms, simply as manifestations of irrationality.

My first task appears therefore to present to the reader what I see to be the social rationality of such counter-movements. This is what I have attempted to do in the first section of this introductory essay. Since it would indeed necessitate a research project to speak with some authority, I have to content myself with asking questions which I regard as pertinent. The main difficulty in describing such counter-movements is that they come into official existence only, so to speak, when science recognizes and defines them as such. I was reminded at times of historical writings which for centuries managed with complete ease and considerable arrogance to ignore the existence of almost 95% of humanity, referring to them only when absolutely necessary, as the lower population, the people or the masses. The hegemony of scientific rationality made even the recognition of the social existence of these counter-movements dependent on their being attacked.

My guidelines in raising the questions contained in the next section have been the different forms of antagonistic interaction between the counter-movements and the sciences. The reaction of the defenders of science has itself been beset by tensions created by the changing social institution of science and its members' collective representations thereof. I reach the conclusion that there is indeed a dormant potential for conflict, still hidden behind the successes of science, which comes to a clash in ideology, social values and objectives in various encounters with the counter-movements. But the question is not whether science should be abolished or not, or whether its rationality can sustain itself. Rather, the social rationality of the counter-movements is one which turns around the question of science for whom?

In Section 2, I have followed those manifestations of the counter-movements which are most coherent and perhaps, therefore, also most visible. I conceive of them as forming a continuum between anti-science and science.

At one end, incompatability is at its maximum, extending into problems of how knowledge is gained and verified as well as into sharply divergent social aims.

At the other end, overlap is considerable but there are divergencies on the legitimacy of different forms of criticism and on the boundaries of what constitute legitimate scientific problems as compared to political (non-scientific) ones. In singling out three manifestations of counter-movements, I have followed to some extent an empirical clustering which, however, is always much more scattered and overlapping in reality that any neat analytic representation would like to make believe. In doing so, I have retained the common terminology of anti-science in the romantic tradition, pseudo-science and critical science, and have abstained deliberately from an attempt to redefine them by using criteria of scientific rationality. I have sought instead to bring out the social basis of these movements from which, from a sociological point of view, their cognitive strategies can be deduced. In the course of this procedure, they may have gained more respectability than some readers are perhaps willing to grant to them. Because the counter-movements are documented via the commentaries of scientists, I have always been forced to look across the fence at the same time and some curious parallels and mirror-images have emerged between science and the counter-movements, greatly reinforcing my initial suspicion that we are dealing with two sides of a single social process. In its course, domination produces subordination, and subordination various forms of rebellion and resistance, whereby the assumptions on which domination are placed, are challenged and eventually perhaps also changed.

I have been brief on critical science, not because I consider it less important, but because critical voices, which come largely from inside science, can safely be left to speak for themselves. It has not been my intention to write this essay around the papers brought together in this volume, although the experience of collecting them (and the ones which we were unable to obtain) is certainly reflected. Rather, my intention has been to provide a sociological introduction to a field of conflict between science and its critics which contains, in my opinion, rich material for thought. Finally, I have refrained from engaging in speculations about the innovative potential of counter-movements: it is up to the reader to judge on the basic of what is contained in this volume and what he or she can observe happening outside it.

2. Rationality: Scientific or Social?

To most readers, science and anti-science will appear as two clearly separate phenomena with sharp epistemological and institutional boundaries existing between them. They will depend on one's conception of what science is and ought to be. The place accorded to rationality as the prime ordering force in thought and human affairs, and one's sympathies with various critics of science, ranging from Blake to the Bulletin of the Atomic Scientists, will no longer let appear the actual boundaries between what constitutes science and what anti-science as fixed and immutable as they were at first sight. Add to this most scientists' reluctance "to volunteer for a debate for which they neither claim particular expertise nor expect much reward (since) life is short and research long" (1), and the current flourishing market for the so-called pseudo-sciences, and you will find a contemporary scene as rich in expressions of reservations against official scientific practices and their consequences, as it is bewildering in conflicting claims and counter-claims. To make matters more confusing, audiences to which attacks and warnings are addressed, whether of potential adherents and present supporters, tend also to vary. Although there is some overlap, it is difficult to gauge their impact on what, in the spirit of blissful ignorance, is usually called the public. Anti- and pseudo-scientific practices and ideologies have attained uneven social visibility, further exacerbated by the fluctuations inherent in any social movement. This renders it extremely difficult to separate what is actually happening from what is perceived to happen, actual threat from perceived threat which, in accordance with an old sociological wisdom, is, however, no less real in its consequences.

The scene of the hotly disputed territory over what belongs to the realm of rationality, reason and science, and what to the opposing forces of darkness, unreason and disorder, is difficult to survey. Language is treacherous. Labels are applied freely, with little regard for their analytic value, since their main use is instrumental in an ongoing antagonistic exchange. Terms like anti-science and pseudo-science still reveal, in the context in which they are used, the original intention of attacking or ridiculing an opponent (2). Other, even more explicit, forms of verbal injury provide clues as to where the scientific establishment feels threatened or proves vulnerable. When responding to what they often see as attacks on their personal integrity and their work as scientists as well as on the institution of science, scientists react with unusual passion and public display of emotion. White some may draw comfort from reading the figures of the latest opinion polls, which reassuringly show that the often proclaimed crisis of science is far from reaching alarming proportions (3), others can be found organizing self-help groups — for instance, the recently founded Committee for the Scientific Investigation of Claims of the Paranormal. As happens frequently in situations characterized by social as well as cognitive uncertainty, unverifiable rumours are spread. Their very nature lies in a plausibility which escapes verificability, making for a somewhat bizarre definition of social reality. One of the more persistent rumours draws on parallels between the present situation and a variety of historical predecessors (4). Difficult to prove, since neither historical situations nor actors are ever completely identical, yet appealing in emotional content, such analogies produce a rich and evocative social imagery, recalled by the equally evocative parallels which are said to exist between Modern Physics and Eastern mysticism (5).

Any attempt to apply a critical sociological analysis to the conflicting claims about the social significance of science is faced with difficulties which stem partly from the ideological nature of the conflict, partly from the factual elusiveness inherent to social movements. For the sociologist, the question is not how much irrationality emerges from the counter-movements, but how to probe their social rationality. What dimensions of science and its claims are being attacked? Are these attacks real, perceived or only imagined, when judging from the responses of scientists? (6).

At this stage, I can only outline some questions which indicate the range still open to detailed investigation.

Social rationality provides an account of world views, beliefs, and ideologies of groups on the basis of a specific social and economic situation and in a concrete historical context. It starts from the assumption that men and women have valid social reasons to hold the ideas and beliefs they do and, by collectively endorsing them, to refer to them as knowledge. This is regardless of how they may be judged by another group, in other historical circumstances or on the basis of other forms of knowledge. Since we are accumstomed, however, to assign to scientific knowledge a superior social, as well as epistemological status, to which we accord the privilege to judge what others believe as right or wrong, it is all too easy to dismiss as irrational, emotional, and unfounded, everything falling short of the standards of scientific rationality. In assuming that others' — social — standards are equally valid, the sociologist is looking at science as a social institution and at scientific knowledge as being equally socially constructed. Seen from this perspective, the conflict between science and anti-science is about different claims to define standards of rationality with universal validity, and to subsequently judge actions and set social objectives accordingly. One of the first steps in probing into the social rationality of counter-movements is therefore to reveal the symmetry between science and anti-science. This can be done by analyzing the social rationality which governs scientists and leads them to believe and respond to what they define as anti-scientific irrationality.

2.1. *The Social Rationality of the Defenders of Science*

The habitual disregard of anti-science and pseudo-science (not conceived as parallel communities) by the major part of the scientific community is punctuated at times by outbursts of worries and hostility. Under what social conditions do they occur? When do scientists feel sufficiently threatened to respond to the criticisms or practices of an opponent who — at other times — can simply and safely be ignored as inferior and harmless? What is the political pressure behind scientists' periodic attempts to convince the public and politicians that continuous financial and public support is needed and, when is it judged opportune, to link it with a warning that the rising wave of anti-scientific feelings has to be halted? Upon re-reading some of the publicly voiced concerns of the early 1970's, one may gain the impression that the best defence strategy of scientists has been to attack, with the counter-culture of that time as a ready and obvious object. What have been their genuine worries, hidden perhaps behind the motive to ensure continuous support on the institutional and recruitment level? (7). What, may we ask, in the present debates on the controversial applications of science and technology, recombinant DNA and public participation, is due to fear of loss of professional autonomy? How easy is it to yield to the well-entrenched stereotype of an emotional public, lacking objectivity and understanding of the real issues (8) and to use it as a pretext only in a politically opportune way?

To question the motivations of scientists is not to belittle their genuine worries and fears, but simply to show that they too have interests which may be at stake. Social rationality is at work in defence, just as it can be seen to operate on the side of the critics.

In order to proceed further along this line of inquiry, we ought to turn to the scientific community as a whole and ask who it is who responds to attacks on science and (perhaps) their own work. Who are the official or self-appointed spokesmen writing in the editorials and readers' columns of widely circulating scientific journals? What is the image of science which they rally to defend, and how does it compare with the actual working conditions they themselves, as well as most other scientists, find themselves in?

Impressionistically, I have detected a certain amount of nostalgia in the reactions to anti-science. A bygone state of science is often evoked, and ideals are being upheld which no longer seem rooted in social reality. This is especially so with regard to an often heard remedy. Everything would be well with science and its relations to society, the argument goes, if only present ties which bind science too closely to the corrupting influence of politics and power could be cut loose (9). By separating off criticisms of the 'merely technical relations' between science and society, that is, those directed against applications, science could be protected from criticism and guard its purity. To me, this is a naïve wish, but understandable as the strategy of an argument based on social rationality.

In other instances in which opponents are attacked, the underlying purpose seems to be to disavow potential competitors. They are charged with not doing science at all when compared to an implicit, but variable, standard of quality science. How are we to distinguish the socially trivial reaction of a conservative scientist who shrinks back from the idea of new and unconventional knowledge being introduced, from the non-trivial response to new paradigms? When are the boundaries of what constitutes 'real science' defended as immutable by disciplinary or institutionally vested interests and when do we witness the social process of setting new scientific standards and drawing new boundaries? Two separate phenomena need to be dealt with in this connection. One is the well-known problem of the delayed acceptance of new kinds of knowledge whose theoretical or methodological significance was not recognized. Often ridiculed and rejected, with occasionally tragic consequences for the individual involved (10), they reveal the almost pathological abhorrence many scientists feel when confronted with charlatans. The need for reassuring and precise criteria for demarcation is very pronounced and rigorously enforced against outsiders. The cognitive and social mechanisms of exclusion which are brought into play here contrast sharply with the openness towards the possibility of errors which is professed at other times. The other problem concerns the difficulties of separating legitimate criticism of the uses and abuses of science from criticism which is judged to go 'too far' (11). Conflicts over boundaries between what is considered to be still within the realm of legitimate scientific concern and what transcends it usually hide social and political interests and are part of a frequently practiced strategy.

With this symmetry in mind, let us turn our attention now to what constitutes anti-science and to what it is precisely that scientists respond. The intellectual core of the irrationalities attributed to the critics which figure so prominently in the reactions of scientists are secondary from a sociological point of view insofar as they need to be related to the social basis.

2.2. *The Social Visibility of the Counter-movements*

Most of the generalizations, despite some historical research (12), consist of impressionistic descriptions rather than serious social analysis. This has been facilitated by focussing on the public declarations made by the heroes of the counter-culture, for instance, or by concentrating attention on those champions of the pseudo-sciences who have reached a large lay audience via the mass media. Until recently, and with few exceptions, the sociology of science has been little concerned with rival claims to scientific knowledge (13), leaving it to cultural critics and the mass media to offer explanations for marketable manifestations, like the burgeoning cults of zen and other forms of Eastern mysticism, occult practices and the commercial exploitation of various encounters of the third kind (14).

It is not an easy task to gain access to a dispersed and on the whole unorganized counter-movement (15). Obscure little magazines purport to speak for an indeterminate dissenting majority, while the lore of an 'alternative science', carried by the optimistic exaggerations which stem from the desire for a new and integrated life style, projects visions of an ecologically more viable and happier alternative future whose basis is difficult to assess. To this, one must add the differing social visibility of various parts of these counter-movements. While some of its proponents have gained wide publicity (16), others appear to work, hardly visible at all, on the fringes of official science. They do, rather than write or speak, and for understandable reasons, show little inclination to become the target of a potential attack. Why, we may ask, are some members of the counter-movements content to cut out for themselves a niche in which they hope to survive, while others have sought and attained public attention and operate with missionary zeal? Is it solely individual talent and a flair for publicity which accounts for such differences, or are there social reasons which may explain the greater or lesser readiness of the mass media and the public to respond to pseudo-scientific promises and hopes? If so, what are the needs and hopes which official science has not been able to satisfy, and what follows from this?

There is, in addition, another manifestation of a counter-movement, which is not easily classified as either anti- or pseudo-science, but fits more into the image of suppressed forms of knowledge and alternative practices, occasionally called folk science. The history of scientific medicine is especially rich in struggles between different social groups, some of whom were simply driven outside the officially recognized boundaries of the self-organizing and victorious profession, with their knowledge and practices being de-valued and officially forgotten (17). In other cases, inter- and intra-professional competition may not have provided the necessary social basis for survival in the form of a counter-movement, but a fund of knowledge and skills has nevertheless been driven out, suppressed or forgotten, in a process of social amnesia. What kind of suppressed knowledge and practices are resistant and under what social conditions may they survive, however precarious and remote from the dominant form of scientific knowledge, and which ones are doomed to share the fate of other heresies?

Social visibility depends upon several factors: the hegemony of scientific rationality in singling out what it defines as heresies; the heterogeneity of the movements themselves, varying in missionary zeal and practice; the receptivity of a society whose needs may not be met by official science, seeking consolation in beliefs which are closer to their concerns, and, finally, the social conditions which create more favourable prospects of survival for some forms of knowledge deviating from scientific rationality, than for others.

2.3. Science for Whom?

Finally, there is the question of the relation between the foregoing concerns and change within science itself. The scientific conversion of the world, as it has been achieved by Western science, has not only added to the ongoing process of secularization of religion, but has led science to take over many features of a religious institution, albeit one devoid of an explicit theology. Science holds the monopoly of truth and ways of achieving it through the process of scientific inquiry. It claims that the major benefits which have been bestowed upon humanity are direct or indirect results of the development of science and its applications through technology. It has set up a highly efficient recruitment system, extremely demanding in single-minded dedication and promising high rewards, a hierarchical structure of forms of knowledge and their practioners inside science. Radical breakthroughs in gaining new knowledge are awaited as eagerly as were previously the miracles worked by certain saints. Due to the peculiarities of the scientific method and the institutional context in which it is being practiced, the ideal has been to separate objectivity from subjectivity, facts from values, science from the rest of social and political life. Quantification has become a method not only of measuring, but of constituting, the objects of scientific inquiry. In doing all this, science has developed its own ideology, suited to underpin and defend its successful practices, culminating in the ultimate proof of its success: science works. It implies the separation of a process of inquiry and its ensuing results, of research and application, of good and 'healthy' science and its abuses and pathologies. It has generated an ideology negating the existence of ideology.

All of this is too well-known to need further elaboration (18). But in the process, science had to change itself, not only in size and in results, but in the process of how knowledge is being produced. It is with regard to some key elements in the changing production process, that the antagonistic relationship between science and anti- and pseudo-science becomes relevant again. Collective representations of what science is, and does, usually lag behind actual changes taking place in the infra-structure. Conflicts challenging the assumptions on which collective representations rest, may bring out into the open the kind of changes that are occurring. What remains of Nature, we may ask, given the state of physics today, when few questions are left which can be worked upon without access to high-powered technological equipment? What, other than partial comprehension, remains for the individual researcher who, as a member of a large team is subject to increasing pressure of intellectual division of labour? The intensification of intellectual labour proceeds at an alarming rate, necessitating ever larger laboratories, more expensive equipment and larger research teams. The increasing abstractness of the symbolic production of knowledge, exemplified by the open-ended search for quarks in physics, leads to an automated interplay between theorists whose hypotheses are in need of experimental set-ups which can no longer be afforded, and results eventually in a concentration on a decreasing number of problems, while the rest have to remain unresearchable. What will happen to experiments, the essence of the scientific method of inquiry, if it becomes no longer practically feasible to produce experimental results (19) and, in some cases, no longer possible to reproduce them, due to prohibitive costs or demands for what amounts to a unique experimental set-up due to costs (20)? If it is true that science is on its way to making symbolic production its dominant mode of producing knowledge, what will be the consequence?

Is everything which is scientifically possible to be realized and put into practice? Who is to decide? Who will control not only the general direction in which science is moving, but whether a particular research programme, a promising line of inquiry, is to be pursued further or not? While current debate and periodically emerging controversies forcefully attract the attention of scientists and politicians to these issues, they ought to be seen in a wider context. If, in the whole history of humanity we have never been in the possession of so much and so powerful knowledge — what are we to do with it? How much of it, and which part of it, is to be put at the disposal of those who are already powerful, and what is there left for the others? And although most scientists are genuinely convinced of the social relevance of their work, at least through some indirect benefits which flow from science to the people, few of them ever put themselves in the position of those who are at the receiving end, whose world-view and life situation is a long way from what science would lead one to believe. This is not just a matter of choosing the right policies for science education, or seeking to improve communication between the scientific community and the lay public, but a fundamental question: Science for Whom?

The direction in which science develops makes this the central question, which has occupied a prominent place in the discussions of the counter-movements.

3. The Social Rationality of the Counter-movements

As already mentioned, any attempt to define anti-science or to distinguish its different manifestations which uses the demarcation criteria of scientific rationality is bound to repeat the accusation of irrationality without elucidating the existing social relations between science and anti-science. The conflict between them is at different levels. While part of it takes place on an epistemological level, another part opposes social groups who are the carriers of claims and counter-claims. Differing social values and objectives can make for mutual incompatibilities and total confrontation, while, in other cases, disagreement may only be partial. Seen from such a perspective, science and anti-science are not two separate entities, but the two ends of an uneven continuum.

Most prominent on the far end of anti-science, the denial of scientific rationality has taken on a persistent form which has commonly been called the romantic, Utopian or literary-humanistic, tradition. Reaching back to modes of thought that have preceded the rise of science, sharpened by the ideological and social battle between Enlightenment and Romanticism, reinforced and enriched by literary creative visions, this tradition forms an almost classical antithesis to the equally stereotypical image of science as embodying cold rationality and, to a certain degree, inhumanity. Holistic against particulate or atomistic, subjective against objective, quality against quantity, synthetic against analytic, emotional against rational, traditional or reactionary against progressive — are some of the better known dichotomies which have been used to characterize the differences between these 'styles of thought' (21). The latest representatives of this tradition have been the proponents of the so-called counter-culture and at least some branches of the ecological movement.

While science has subordinated nature epistemologically, in its theories and in its practice, the marked difference is that for many counter-movements nature is granted an existence independent of science. It is 'out there'; men and women still relate to it or are conceived to be part of it. Their social relations are embedded in it. They challenge the monopoly that scientific rationality claims to possess over reality — claiming instead that different kinds of realities exist, not all of which can be approached through scientific experience alone. Counter-movements differ in the extent to which they refuse to subsume nature — or society — under the exclusive dominance of science, but, for all of them, important consequences for valuing different forms of knowledge and for striving towards different social objectives follow.

3.1. The Romantic Tradition of Anti-science

Disagreements over the value accorded to nature and to the relation between humanity and nature is perhaps most marked in the romantic tradition. Nature is something not to be tampered with. The equilibrium in the mutual relationship between humans and nature is a precarious one, ideally a harmonious one. Where science is seen as closely associated with the ongoing transformation of nature, even its de-naturalization in the sense of scientific reductionism and increasing interference with a variety of natural processes, the romantic tradition prescribes non-interference. Let nature do her own thing and let us act likewise, has been an underlying theme expressed both in the exuberent self-confidence of the counter-culture at its height, and of the ecological movement in a more apocalyptic mood (22). The clash in social values and prescriptions for social action which are derived from them, results in total incompatibility of views and actions. Since science has been the prime instrument in the transformation of nature, all attempts to stop this process have to be directed against science as well.

Social analysts and historians have drawn attention to the social roots of such values. They have been identified in those social strata whose previously dominant position in society has either been undermined or is threatened by the changes generated through industrialization and the processes of modernization (23). The reaction of the romantics against interference with nature can consequently be interpreted as a reaction against interference with their own social position, or whatever power and status they have previously held. The romantics cannot wish for change, since it is their position and they themselves who will be the first victims of change. They are opposed to active intervention on the part of science and technology, since their social position, tied by ideology and practice to the present state of nature, is being undermined. Today, the task is no longer to defend a natural social order, which has often, though not always, been linked to the more conservative forces in society. Rather, what is being defended is seen to be the last realm of subjective freedom and autonomy, escape from further regulation and standardization imposed in the name of scientific and technological progress. It is these interests which find expression in the desire for non-interference, in the emphatic battle cry of 'doing your own thing'.

Critics of the romantic tradition, with varying degrees of sympathy and understanding for the underlying theme of opposition, usually point to the unreal and Utopian elements of the Arcadian dream. The verdict is often a very harsh one (24). But the uneven battle, on the level of social and (muted) political action, is complemented by one on the epistemological level. The priority accorded by the romantic tradition to the subjective side of gaining knowledge, verifying and certifying it, does not constitute an arbitrary or irrational claim for subjectivity as a valid method of producing knowledge. Rather, the claim is that knowledge thus produced is in itself a sufficient basis for the social production of knowledge. Few scientists would deny that there is room for individual creativity and even spontaneity in the generation of insights and theories. They have to deny, however, that subjectively gained insights can aspire to the status of socially produced and acceptable knowledge. The essential disagreement arises over conflicting claims as to how knowledge which is gained subjectively is to be verified and certified. Science claims that this has to be done in accordance with the canons of objectivity, as remote and free from intrusion of the subject as possible, while the romantic tradition would contend that socially acceptable knowledge can be produced by mechanisms of subjective verification and certification, as long as these are socially shared mechanisms. It is obvious that this would eliminate the ways of doing science as it is practised today. It presupposes a society which has highly developed mechanisms for sharing and verifying subjectively gained knowledge, a society which appeals to organic, rather than mechanistic, solidarity, and which is sufficiently unified in its collective representations to work through the 'natural' consensus of subjects rather than through objective and mechanical instruments.

This is not the place to assess how much subjectivity has been hiding under the cloak of scientific objectivity. What we may conclude is that the romantic tradition opposes science as the enemy of nature of which the romantics wish to be part. It is their place in society which is at stake. Only non-interference will guarantee that their social interests can be preserved, that they will not become the victims of changes brought about by science. The corresponding direct access to nature constitutes for them the social basis for producing knowledge. It is valid knowledge since it promises to preserve the relations between nature and a society in which they can claim to be part of it, but from which science, in its present form, will be excluded.

3.2. The Pseudo-sciences

Further along the lines of the continuum stretching between anti-science and science, we encounter a cluster of practices and ideologies usually referred to as pseudo-science, whose more respectable branches have succeeded in calling themselves by the more neutral term para-sciences. A different kind of threat is posed to science from these quarters of rebellion. Compared to the manifestations of anti-science grounded in a romantic world view, the incompatibilities between science and the pseudo-sciences are more partial than total. In many ways, the pseudo-sciences aspire to become scientific. They imitate not only selected features of the official ways of doing science, but they can also be seen to search for a new synthesis with science. Thus, for example, future prospects for a great breakthrough in Western science and in the occult sciences of the Orient are said to be good (25). Astrology will again become a recognized science, once it has made use of cybernetics and statistical analysis (26). But such claims still fall on the deaf ears of official science.

The fundamental basis for contention is the question what constitutes a phenomenon worthy of scientific investigation and what kind of science is to be practiced in the course of doing so. The reproach levelled against official science from the pseudo-sciences is that the former is neglecting the investigation of certain phenomena which are part of nature, but which are arbitrarily dismissed by science as non-existent, unscientific or pseudo-scientific. But what lies behind such divergent cognitive and methodological practices of inclusion and exclusion? The diversity of lines of inquiry, objects of investigation and methods used in explaining them within the wide field of pseudo-science is staggering and makes any attempt at generalization appear quite futile. Nevertheless, certain features stand out more prominently than others. Behind the ardent pursuit of officially neglected or denied phenomena lies the hope and the need to prove their existence. These may be spiritual or psychic phenomena for which there is no room in official science, holistic versions whose counterparts are hardly recognizable in the atomized, mechanized elements, scattered over scientific disciplines and reduced in the name of science to a purely 'materialistic' or physiological basis. Other phenomena are to be found on the fringes of scientific disciplines, as in para-medicine and para-psychology, or beyond the frontier, moving about freely in extra-terrestrial spheres.

The persistent search for phenomena that have been excluded from official science is sustained by the hope and the confidence that there is still room for single-handed discoveries. It is socially backed by the claim that access to nature should still be possible without the interference of intervening high-powered technical apparatus which official science has effectively monopolized for its own practioners, thus serving as a powerful mechanism for the exclusion of all who are not part of the scientific guild. With it goes a refusal to recognize the apparently arbitrary official disciplinary boundaries and subfields of research specialities, backed only by the institutionalized vested interests of those who control scientific and technological equipment, and with them the methods deemed appropriate for scientific inquiry.

It is not a 'natural' affinity for holism, nor primarily the attraction for the occult or other spiritual phenomena as such which can account for the form the conflict between official and pseudo-science takes. Rather, we have to focus first on science as having effectively monopolized two different claims. Firstly, it holds a monopoly on the totality of all phenomena occurring in nature (and, by extension, in humanity) which are known already at present, but also those that may still be discovered in the future. Part of this monopolistic claim to define what constitutes a phenomenon and how it may be investigated, extends also to the increasing intellectual division of labour between, and within, disciplines. Secondly, science holds an increasingly powerful monopoly of access to any investigation of nature. The scientific method has resulted in measurement devices and technical instruments for observation and experimentation which render any form of direct access to nature impossible, reducing it to what must appear as utterly simplistic and naive attempts of dilettantism. The mechanisms for social exclusion erected by science are therefore effective in two ways, epistemologically and socially. In their counter-strategy, the adherents of the pseudo-sciences make two counter-claims. They contend, first, that the monopoly of science over the definition of nature is a pseudo-claim, that, on the contrary, phenomena can be proven to exist in nature which science has persistently denied or neglected. By cognitive necessity, these phenomena are construed in holistic, rather than reductionist fashion, cutting across the artificial disciplinary boundaries. The second counter-claim contends that — however powerful may be the intervening technical equipment for observation and experimentation which science has set up to control access to nature — direct access is still possible. That this is so, is proven in often ingenious ways, by telepathy, the systematic observation of UFO's and other phenomena without the aid of technology, or by setting up measuring devices which use low technology and in some ways are reminiscent of the early days of science, when there was still room for the individual amateur. Driven underground or out to the fringes, having become obsolete through the advances of scientific and technological development, the pseudo-sciences represent a kind of do-it-yourself science. It is their social and cognitive exclusion which accounts for the turning towards those phenomena which science denies and to which they can still claim to have direct access in their pursuit of knowing nature, rather than the other way around.

3.3. On Social Relevance of Non-scientific Truth

The occult sciences and their like, as they have recently emerged from the underground, seem to promise a rich field of activity for all those would-be scientists and amateurs who, long ago, had to give up their last foothold in science as it became industrialized. An increasingly esoteric and elitist 'high science' has nothing but contempt for them (27). The 'low sciences', i.e. folk and pseudo-science, have become the obvious recruiting ground, for all those who fight the scientific monopoly, to define what the world is and what is in it. They stand for the right to other kinds of knowledge than scientific knowledge as defined officially, and for a future in which, in the words of Levy-Leblond, we may have to recognize one day that, for knowledge to be true and just, it does not necessarily mean that it has also to be scientific (28). But, however one may be tempted to celebrate the popular defiance against a science turned elitist, there are clear signs of the vulnerability of these movements to being incorporated by what they wish to oppose.

Until recently, the proponents of anti-science in the romantic tradition have, by and large, not developed any practical alternative to the science they have attacked, while the practioners of the pseudo-sciences (with the exception perhaps of some occult sciences) have been deficient in constructing an alternative to the ideology of official science. The anti-scientists are often attacked with the argument that after all, they too drive cars and use electricity, an argument which has done little to soften their ideological resistance to science. The pseudo-sciences do not possess a common ideology or shared view of society at odds with that of official science. To the extent that official science is ready to open up and to draw new boundaries, pseudo-scientific practices and phenomena may eventually be admitted within it. History shows that many wild blossoms on the fringes have been successfully turned into domesticated and useful species. To the extent that the outsiders show a willingness to submit, in exchange for the often secretly desired prize of official recognition, to the intellectual and organizational division of labour, to relinquish their claim to holism and to adopt technical paraphernalia, they too are on the way to incorporation. In the course of this process, official science may have to change its standards of exactness and scientificity, but it may be a price worth paying for gaining new and ardent believers.

Incorporation or innovation are contrary hopes or fears. Only recently, and to an extent which is difficult to gauge, has there been some overlap between practitioners and ideologues, raising hopes for some kind of an 'alternative science'. Most of it, I would suspect, is confined to little corners in which do-it-yourself science can be practised, guided by a value orientation sufficiently strong to sustain it (29). In order to turn into a serious challenge, it would need to invade the laboratories of official science and to challenge the scientific ideology in the heads of practising scientists. For social innovation, in the sense of an alternative, to occur, it is neither sufficient just to preach, nor to do without preaching. For new cosmologies to arise and for innovation to take place which can interpret the world anew and transform it accordingly, ideology and practice need to coincide.

But social relevance or lack thereof are not sufficient conditions to bring about changes, certainly not in the short-run. They remind us, however, to look once more at the interests which are being served by different kinds of knowledge.

3.4. Mirror-images of Science and Anti-science

It may be tempting to interpret anti-science and pseudo-science alike as movements which are out of tune with social reality, as a reaction on the part of those who have been excluded from participating to the full extent that they may have thought to have been entitled. In the case of anti-science, the excluded social groups can be seen as comprising all those who had to incur losses in terms of their own social status or privileges and who continue to cling to a value system which once provided the social basis for their position in society. In the process of industrialization and related developments aided by science, their standing has been threatened or undermined by new and upward mobile social groups who had to gain from propagating science. An explanation along these lines has, in fact, been the dominant one with regard to all those who are or have been discontent with industrialization and its effects and have turned against science for the part it has played in bringing it about. In the case of the pseudo-sciences, we could follow a similar, although more limited explanation. The public enchantment with the occult and other pseudo-scientific preoccupations may be said to stem from a vacuum, created by the neglect with which official science has treated the spiritual needs of its laity, who therefore become an easy prey of heresies. The social exclusion of the practitioners of the pseudo-sciences goes a long way to explain why they had to turn to a different cognitive understanding of nature and to develop their own alternative practices of gaining access to it.

But in looking at the other side of the coin we could also maintain that it is the defence of science which is out of tune with social reality. There is also a romantic tradition inside science which feels called upon to defend an ideal of science which no longer corresponds to present-day reality — and indeed is doubtful whether it ever did. It is the ideal of 'pure' science, untainted and untouched by the demands being made upon it by the political system, which is being upheld as the only route open for science to salvation. "The only candid answer" to the question why the makers of the counter-culture automatically regarded science as being the enemy is seen residing in the fact that "over the last 30—40 years the profession of science has lost its public reputation for living up to its proper ideals" (30). By locating the anti-scientific mood of the public in its perception of science as a political force, stemming from its too close association with power and authority, we are being sensitized for the necessary reforms. The shortcomings, in the scientific-romantic view, spring from political, not scientific or technical, defects in the institutions in which science is embedded. Problems over which public concern has been rising, are said to be 'only political ones' (31) and implied is the pressure for a neat separation between politics and science. Likewise, the moral appeal to the traditional scientific value system of an uncorrupted, disinterested science, implies that virtue can be restored, if only scientists were to cut their ties with the political and military establishment and return to what their anti-scientific counterpart would call 'doing their own thing'.

But also those who defend science against the aspiring pseudo-sciences reveal in their response the insistence on criteria of demarcation which are slowly, but surely, giving way. Not only are the attempts to delineate once and for all what is scientific from non- or pseudo-scientific faltering, if not on the epistemological impossibility (32), then on historical relativism (33). Within frontier science, for instance, the demarcation becomes completely impossible (34) and Levy-Leblond has recently questioned the claim of exactness of the natural sciences and has proposed to look instead at physics as a social science. Even the most rational of the sciences is infiltrated every day by a multitude of irrational elements. The delineation of a science has to proceed by expropriation and exclusion on an institutional basis, which is less an epistemological than a political operation. The certainty of physics represents consequently only a superficial criterion of demarcation: behind the certitude of this reality lies another kind of reality, like passing trains, where one may hide the other (35).

The question then becomes: which reality do we want?

The point simply is, that science, in the course of establishing its hegemonic dominance over other forms of knowledge has changed itself. So have the epistemological preconditions, the place accorded to nature and access to it, as well as the social and political context in which the work of scientists is organized. A return to the idyllic state of a science free from external constraints and demands is as Utopian as the counter-culture's fantasies of "trying to tame (the) generals with telepathic messages and the thought (that) the greatest concentration of corporate wealth the world has ever seen (could be humanized) by going barefoot and eating unhomogenized peanut butter" (36). The separation of the applications of science and its technological offspring in its political use from the process of production of scientific knowledge itself is as illusory as life without cars and electricity. The Appolonians of science, i.e. its philosophers (37), who have appointed themselves self-righteously as eloquent defenders of the Faith (of rationality), may have to learn to live with the public scandal of the new alliance between scientific rationality and some irrational, but socially more significant, pseudo-scientific practices. The attacks on science as well as its defence have to be seen as dialectical counterparts. Science has increasingly imposed its rule on the world, but in doing so has become subject to the workings of the same social and political mechanisms which it had hoped to escape via scientific objectivity.

3.5. Critical Science

Some of them are catching up rapidly, opening new lines for criticism of science which can only be criefly touched here. The spectre of global extinction, ecological collapse or uninhibited gene manipulation makes some measure of public control and accountability on the part of scientists appear mandatory. But how far should control be extended and how should accountability be exercised? What kind of criticism is still legitimate and when is it warded off as emotional overreaction or latent hostility from the laity? How far should external influences, political in the widest sense, extend into scientific activities? These are only some of the disagreements with which the last group of critics on our continuum between science and anti-science has to contend. Coming largely from within, critical science turns against those aspects which it regards as irresponsible on a variety of humanistic, moral and political grounds (38). Critics oppose a view of omnipotent science which claims legitimacy to indiscriminately bestow upon humanity whatever passes through the heads of scientists, regardless of the consequences. What separates critics and staunch defenders of a hard-line science is a shifting definition of legitimacy. While one side claims what, in German, is called 'Machbarkeit', the right as well as the capability to do everything which science renders possible, the other side argues for moral, humanistic or political criteria of choice. One one hand, it is society which is expected to adapt or be adapted to the world science and technology have helped to create, while on the other hand, at least a certain measure of adaptability to social needs and structures is urged on science and technology.

Critical science has been united by issues where an underlying political consensus has existed, the opposition against the war in Vietnam being the most spectacular example. But consensus begins to break down, when issues of applicability of particular research results arise, of assessment of different forms of risks or in discussing particular forms of control. All too rapidly they tend to become entangled with the self-interest of those whose own work is incriminated or the collective interests at stake when professional autonomy is threatened by external regulations. While granting to critics the right to speak up against abuses, too much criticism, especially when uttered in public, is easily held to be detrimental to science and attempts will be made to contain it in the proper channels for such communication, namely inside the scientific community of interest (39). Although most scientists would willingly share ecological concerns in the abstract, only a minority has become actively engaged in the public debates surrounding nuclear energy and related issues. And while politically sensitized and committed scientists venture to analyze science and its activities as part of an encompassing and permeating political reality, most of those working in the 'hard sciences' would still wish to draw a sharp and clear line between their own scientific work qua scientific activity which they would claim as exempt from the kind of political analysis they properly apply to the institutional context of their work (40).

But it is doubtful whether, in the long run, such a distinction can be maintained and whether the cognitive content even of a discipline like mathematics can be spared a critical analysis touching its very foundations (41). By subjecting the production of scientific knowledge, and not just its product, to a criticism which links the development of scientific thought to the power structure of a society, the production of this kind of knowledge is being relativized by laying bare its social basis. In the meantime, the defenders of the status quo of science will muster their battalions for the defence of what they must regard as the ultimate step of subversion, inherently more dangerous than any previous manifestation of anti-science, because it comes from within. While the critics would indignantly — and rightly — object to being called anti-scientific, this is nevertheless where and how the decisive battles will be fought. Present skirmishes on issues of applicability of new research results and control, on social assessment procedures and public participation merely hide the more decisive issues of equating scientific knowledge with other, and perhaps new, forms of knowledge, by asking which kind of knowledge do we want and need. Some voices pointing in that direction, in what to me appear prophetic warnings, have been included in this volume. To some readers they may appear as vanguards of subversion, as omnious signs of things to come, of anti-science raising its head again. But whose science and whose anti-science will it be this time?

Notes and References

0. I have carried out a sociological study of experts, both pro and contra, who participated in the information campaign on nuclear energy, organized by the Austrian Government. The scientists debated publicly and had to face a suspicious and occasionally hostile crowd. See Helga Nowotny, *Kernenergie-Gefahr oder Notwendigkeit*, Frankfurt: Suhrkamp 1979.
1. Gerald Holton, "Dionysians, Appolonians, and the Scientific Imagination'. In *The Scientific Imagination: Case Studies*, Cambridge University Press, Cambridge, 1978, pp. 84–110, which analyzes the intellectual core of one form of anti-science from a thematic point of view.
2. We shall continue to use these two terms in the most innocent and neutral sense possible, leaving it to the developing argument to reveal the varying social context and meaning in which they appear.
3. Allan Mazur, 'Public Confidence in Science', *Social Studies of Science*, 7,1,1977. 123–25. The Commission of the European Communities, *Science and European Public Opinion*, Brussels 1977, concludes that "the general public sees science as a central factor in the improvement of daily life and would readily endorse the statement of Francis Bacon at the end of the 16th century: The true and lawful goal of the sciences is none other than this: that human life be enriched with new discoveries and powers", p. 85. See also the section on 'Public Attitudes Towards Science and Technology, in the latest report on *Science Indicators 1976*, issued by the US National Science Foundation, which reassuringly states: "The public continues to have an overwhelmingly positive general reaction to science and technology. Out of four possible evaluations of science and technology, over 70 per cent of the public chose favourable replies in both 1972 and 1976".
4. See, for instance, Benjamin Nelson, 'Prêtres, prophètes, machines, futurs: 1202, 1848, 1984, 2001'. In H. Cavanna (Ed.), *Les Terreurs de l'An 2000*, Hachette, Paris, 1976, pp. 227–246.

    For a detailed discussion of the anti-science movement see Steven Cotgrove, 'Anti-science', *New Scientist*, 12 July, 1973; 'Technology, Rationality and Domination', *Social Studies of Science* 5, 1975; as well as the now already classical studies by Fritz Ringer: *The Decline of the German Mandarins*, Harvard University Press, Cambridge, Mass. 1969; and Joseph Haberer, *Politics and the Community of Science*, Van Nostrand, New York, 1969. For a more recent attempt to draw parallels between the German youth movement and the counter-culture, see John de Graaf, 'The Dangers of the Counterculture', *Undercurrents* 21, 1977.
5. Sal Restivo, 'Parallels and Paradoxes in Modern Physics and Eastern Mysticism: A

Critical Reconnaissance', forthcoming.
6. For an illuminating discussion which took place in the early 1970's see Ciba Foundation Symposium, *Civilization and Science*, Elsevier, Amsterdan, 1972. Also, J. Ronayne, 'Anti-Science and the Politicisation of Scientists', *Australian and New Zealand Journal of Sociology* 12, 2, 1976, pp. 219–235.
7. The concerns of leading scientists about the ability of science to survive the attacks launched against it in the late 1960's and early 1970's has led to an interesting collection of papers, examining the depth and nature of the ambivalent relationship between science and its publics. See Gerald Holton and William A. Blanpied (Eds.) *Science and its Publics: The Changing Relationship*, D. Reidel, Dordrecht, Holland, 1976.
8. Upon reviewing the favourable results of the public attitudes towards science and technology reported in *Science Indicators*, Dan Greenberg concluded:

These findings do invite curiosity as to why so much nonsense continues to come forth about public hostility to science and technology. A good deal of it, I suspect, emanates from the comfort that scientists find in believing that they are beset by powerful irrational forces – that makes it easy for them to explain away difficulties. Another source of the nonsense is in deliberate efforts to depict legitimate public concerns – e.g. over nuclear power, supersonic aviation, uncontrolled recombinant DNA research – as hostility to science. They are nothing of the sort, but since it's politically useful to depict rational concern as a product of irrationality, the myth persists that the public has turned against science and technology. Dan Greenberg, 'How Science Stands', *New Scientist*, 23 February, 1978, pp. 523–24.
9. Stephen Toulmin, 'The Historical Background to the Anti-science Movement', *Ciba Foundation Symposium,* op.cit. Note 6, pp. 23–32.
10. The history of suicides and cases of mental breakdowns among scientists for the types of reason suggested above still waits to be written.
11. A good example of what kind of criticism is accepted and shared and what is rejected as urging 'religious transformation' built on soft ground, is Holton's critique of Mumford's *Myth of the Machine*, Vol. II. *The Pentagon of Power.* See Gerald Holton, 'Lewis Mumford on Science, Technology and Life', in *The Scientific Imagination*, op.cit. Note 1, pp. 255–267.
12. For a good summary, see Steven Cotgrove, 'Styles of Thought', *unpublished paper*, and the literature contained in it.
13. Among the exceptions is the forthcoming volume of the *Sociological Review Monograph Series*, edited by Roy Wallis.
14. The well-known German magazine *Der Spiegel* has recently devoted a major part of its issue to UFO's, the return of superstition and search for extra-terrestrial life. 24 April, 1978, pp. 46–65.
15. Roy Wallis, 'The Moral Career of a Research Project'. In Colin Bell and Howard Newby (Eds.), *Doing Sociological Research*, George Allen and Unwin, London, 1977.

As editors of this volume we had also hoped initially to reach practitioners of 'alternative science', but the obstacles proved too great.
16. Notably Theodore Roszak for the counter-culture movement and Uri Geller for the pseudo-sciences.
17. John Woodward and David Richards (Eds.), *Health Care and Popular Medicine in*

19th Century England, Croom Helm, London, 1977.
18. Hilary Rose and Steven Rose (Eds.), *The Political Economy of Science: Ideology of/in the Natural Sciences* and *The Radicalisation of Science*. The Macmillan Press, London, 1976.
19. Gernot Boehme, Wolfgang van den Daele and Wolfgang Krohn, 'Die Finalisierung der Wissenschaft', *Zeitschrift für Soziologie* 2, 1973, pp. 128–44.
20. Giving rise to Alvin Weinberg's notion of 'trans-science'. Alvin Weinberg, 'Science and Trans-Science', *Science,* 177, editorial, 1972, p. 211.
21. Steven Cotgrove, *op.cit.,* Note 12; Stephen Toulmin, *op.cit.* Note 6, and Gerald Holton, *op.cit.* Note 1, provide rich illustrative material.
22. For a critical analysis of the latter see Rudolf Burger, 'Lebensqualität und Warenproduktion – zur Klassenbasis des Ökologismus', *Wirtschaft und Gesellschaft* 1, 4, 1975, pp. 9–32.
23. The classical authority on this is, of course, Karl Mannheim.
24. Marvin Harris, *Cows, Pigs, Wars and Witches,* Fontana-Collins, London 1975, p. 183, puts it as follows:
      A million chanting Reichs and Roszaks affect the advance and spread of science and technology about as much as the chirping of a single vagrant cricket affects the operation of the automated blast furnace.
25. Randall Collins, 'Towards a Modern Science of the Occult', *Consciousness and Culture* 1, 1, 1977, pp. 43–58.
26. Heinz Fidelsberger, *Astrologie 2000 – Struktur einer Wissenschaft von morgen,* Kremayr and Scherian, Wien, 1972.
27. Ingo Grabner and Wolfgang Reiter, 'Guardians at the Frontiers of Science', *This Volume.*
28. Jean-Marc Levy-Leblond, 'La physique: une science sociale', *Etudes et Recherches Interdisciplinaires sur la Science, Bulletin du GERSULP* 1, 1977, pp. 1–16.
29. Sue Peters, 'The Organic Farming Movement and Alternative Agriculture', *This Volume.*
30. Stephen Toulmin, *op.cit.* Note 6, p. 29.
31. Helga Nowotny, 'Scientific Purity and Nuclear Danger'. In: Everett Mendelsohn, Peter Weingart and Richard Whitley (Eds.), *The Social Production of Scientific Knowledge,* Sociology of the Sciences Yearbook, Vol. 1, D. Reidel, Dordrecht, Holland, 1977, pp. 243–264.
32. T.J. Pinch and H.M. Collins, 'Is Anti-Science not-Science?' *This Volume.*
33. Thomas Kuhn, *The Essential Tension: Selected Studies in Scientific Tradition and Change,* University of Chicago Press, Chicago, 1978.
34. Ingo Grabner and Wolfgang Reiter, *op.cit.* Note 27.
35. Jean-Marc Levy-Leblond, *op.cit.* Note 28.
36. Marvin Harris, *op.cit.* Note 24, p. 183.
37. Gerald Holton, *op.cit.* Note 1, referring to certain philosophers of science.
38. J. R. Ravetz, 'Criticisms of Science'. In: Ina Spiegel-Roesing and Derek de Solla Price (Eds.), *Science, Technology and Society,* Sage Publications, London, 1977, pp. 71–92.
39. E. Mendelsohn, D. Nelkin, and P. Weingart (Eds.), *The Social Assessment of Science,* Science Studies Report, No. 13, Universität Bielefeld, 1978.
40. On this ongoing debate see various contributions in the *Radical Science Journal.*

41. Pioneer attempts in the direction of a sociology of mathematics have been undertaken by David Bloor, *Knowledge and Social Imagery,* Routledge and Kegan Paul, London, 1976; Luke Hodgkin 'Politics and Physical Sciences', *Radical Science Journal* 4, 1976, pp. 29—60. See also the Theme issue of *Social Studies of Science* 8, 1, 1978.

ANTI-ESTABLISHMENT SCIENCE IN SOME BRITISH JOURNALS

JEROME RAVETZ
University of Leeds

1. Introduction

There are some people (though not very many as yet) who do not merely criticize the established science but attempt to create a new sort. Here, I will give a brief survey of the leading tendencies of a political cast in this movement as it appears within Britain. I will use journals rather than books as my source of evidence, for these are better indicators of the presence of stable small groups of people who create, and of larger groups who support them financially and otherwise.

In some respects my sample is biassed and may be misleading. First, I confine myself to politically-oriented groupings rather than those based on, and promoting, radically different world-views. Then, by focussing on 'science' rather than some other aspect of commitment or concern, I may exaggerate some differences between these groupings and their neighbours. Both of these problems will be discussed at the end of the essay, with a view to correcting possible distortions in the main account.

The two journals I shall examine in greatest detail are *Undercurrents* and *Science for People*, which represent what might be called 'libertarian' and 'Socialist' Left, respectively. They are both produced by 'collectives' of mainly impoverished volunteers, and rely heavily on contributions from supporters for their articles. There is some overlap in the articles they publish; but in view of their shared radical commitment, surprisingly little. And we shall see that in the matter of 'style', which for the 'newest' Left is very important, their differences are very deep indeed. In particular, the manner in which a grouping copes with the contradictions in its own position can be very illuminating for its total image of the world and its role in it.

2. Some General Reflections on Science and Anti-Science

To place the varieties of radical science in some coherent perspective, I shall invoke, in a rough and provisional fashion, the well-known analogy between contemporary established science, and traditional institutionalized religion. This is not done for rhetorical purposes. By distinguishing among the various functions of 'religion', and observing which are carried over into 'science', we gain a better understanding, both historical and analytical, of the 'anti-science' tendencies we see displayed in our journals. Briefly, then, we may imagine 'religion' performing the following functions: defining a picture of the whole world, particularly as it gives meaning to human existence; influencing the content of, and setting limits on, the legitimate 'public knowledge' of a society; co-operating with the State in the maintenance of social control and the provision of social welfare; defining and supporting morality at the personal level; providing emotional aid and solace; and providing channels for the achievement of spiritual experience. The classic 'warfare between science and theology in Christendom' was explicitly fought on the second issue, and the victory of science was reflected both there and on the first. Further down the list, away from the intellectual aspects, the decline of 'religion' came through a general transformation of society and personal consciousness. Indeed, in retrospect the initiation of the major changes within science, notably the 'revolution' in the century of Galileo and Newton, now seems to owe as much to 'external factors' (including religion) as to successful discoveries within science itself. So the replacement of 'religion' by 'science' has been partial at best. And even where it has occurred, the institutional form of science, and its basically rather precarious ideological position in society, have inhibited the full development of the dogmatic and totalitarian tendencies which are certainly there in it.

This weaker position of institutionalized science, as a successor to religion in defining much of our intellectual culture, has several consequences relevant to our present study of its critics. For one, they are not so easily disposed of by 'administrative means'. The censorship of anti-science work cannot be more severe than that applying in general in a particular nation. Thus Velikovsky's works were published for the American popular market, and his supporters were able to maintain a campaign, ultimately successful, for some scholarly dialogue on his theories.

Hence, critics of science will be tolerated as much as any other social critics, but, on the other hand, the target of their criticism will be less easily defined. As it has developed historically, established science has a *persona* with wonderfully flexible divisions. The twin goals of knowledge and power join in various degrees and manners, in the definition of the goals, immediate and remote, of the disparate parts of the institutions sometimes called 'science'. This confusion enables the maintenance of a very convenient false-consciousness among scientists and their propagandists. One aspect of it may be summed up in the slogan 'Science gets the credit for penicillin; society takes the blame for the Bomb'. In the context of this discussion, the effect is that the 'science' attacked by the critics is outside of what is usually identified as science, or is at best peripheral to it; and the 'science' that they actually practise or advocate is generally even more remote from the activity of science as publicly defined. This reduces their base in the established community of scientists nearly down to the vanishing point. They have so far survived this deprivation; while the scientists are enabled to proceed in their own ways happily unaware of their criticisms. In this respect, the new critics are even more radical than past religious heretics; however abominable their doctrines may have appeared, they were at least recognised as involved in the same sort of enterprise as those who suppressed them. Now, because of its less complete scope, and the mobility of its definitions, science is largely untouched by its critics.

3. Science as Class-biassed

By an historical accident, the socialist fringe group in science bears the terribly respectable title 'British Society for Social Responsibility on Science'. When the group had become thoroughly radical, it decided to adopt a more appropriate name for its journal. It had as a model the American 'Science for the People'; oddly, it omitted the definite article from its own version, so that the title smells of polite uplift rather than of radical populism.

This may not be too important, for the regular contents of the journal make its intended affiliations clear. It should, in fact, be called 'Science for the Workers'. Of course, not many manual workers would derive immediate benefit from its contents. But the BSSRS also produces a 'Hazards Bulletin' and a series of pamphlets on workplace hazards; indeed, its main regular, effective work is in this field. These materials provide clear and competent technical information, much needed by workers and Safety Representatives, in the context of a militant criticism of management and inspectors. The persistent neglect of research into industrial hazards is eloquent evidence of the class bias of established science. And as the movement for health and safety at work grows, BSSRS activists could find themselves at the centre of an exciting development uniting sympathisers among trade-unionists, environmentalists, medical experts and research scientists in a common campaign on hazards. Indeed, a few years of such work which would certainly involve bitter struggles with all established institutions, could develop people with that matured understanding of 'science for (the) people' which is now admitted to be lacking.

In the meantime, the 'Science for People' collective picks up topics and materials wherever it can. It is at its best when a group of volunteers come together to create an issue of the journal out of their common experience. Some years ago they had an illuminating pioneering survey of 'women in science' (1). Most recently, a 'health group collective' prepared an issue (2). This provided an excellent survey of the British Health Service from a radical perspective, ranging from the struggles against the cuts in finance, to those against bureaucratic controls, through to class and sexist biases in medical practice, and even a brief lok at 'alternative' approaches. The previous issue (3) also had medical articles, on dentistry and eyeglasses ('the epidemic everyone ignores'). Another special issue was on 'images of science' (4), where a number of disillusioned scientists describe their experiences in education, research and industry. The futility, boredom and proletarianisation of so much of 'science' is well described. Since the seamy side of the world of science is only rarely studied by academic sociological researchers, this collection of testimonies, however small and unrepresentative, is an important source.

The regular run of articles covers topics in technology, the political uses of science (as torture, XYY chromosome research), and the demystification of the academic end of the enterprise. Only occasionally is there something found to praise: the Lucas shopstewards' initiative; some small co-operative ventures; and science in National Liberation movements in the Third World. The topics left unmentioned included: alternative cosmologies and lifestyles; an objectionable high-technology project that happens to be favoured by the established Left, such as the supersonic transport Concorde; and the abuses of science in (self-styled) Socialist countries.

This last omission brings out a very deep contradiction in the approach of the journal and its sponsoring group. The terms 'capitalism' and 'Socialism' are used freely, one as the basic cause and the other as the only cure for all problems. But neither term is ever defined, even when (as in the case of not-for-profit bureaucracies like the Health Service) the problems will clearly not be obliterated by formal changes in 'ownership'. Unlike many political groups at that end of the spectrum, the BSSRS know that they have a problem. Their most recent Policy Statement admits "There are no easy answers, and no model societies". However, the problem is left there, rather than the phenomena common to high-technology societies being used as the bases for a general radical critique. And in the absence of an awareness of all the moves in the game they are playing, there is a danger that their revolutionary yearnings will be translated only into those radical actions which accelerate certain marginal reforms which the system had already accepted. Safety at work, doubtless an issue where humanitarian impulses and practical militancy combine to good effect, could be just such a campaign whereby the Left is recruited for good works.

Of course, the 'Socialist' Left is not devoid of theoretical speculation. The collective work edited by Hilary Rose and Steven Rose, *The Radicalisation of Science*, and *The Political Economy of Science* (5) is sufficient evidence of a continuing and worthwhile commitment from a broadly Marxist standpoint. As to journals (following my self-imposed terms of reference), there is *Radical Science Journal*, which appears a little more frequently than annually. It features full-length articles on particular problems and abuse, and some extended theoretical essays; the most significant of these recently is 'Science *is* Social Relations' by Bob Young (6). In this, as elsewhere, there is a serious attempt to cope with the contradictions of being a politically radical member of an established intellectual-worker class, but as sympathetic critics observe (see the letter from John Goodman, (7)) the common assumption of a straightforward problem in mobilizing 'the working class' is not realistic.

4. High Technology as Part of a Sick System

The magazine *Undercurrents* began very modestly and soberly in 1972. Its initial Editorial had the title 'Science with a Human Face' (recalling Czechoslovakia, 1968). After reviewing the standard abuses of science and technology, it called for a "sadder but wiser" science, and a technology reoriented to humanity, on a small scale. With only a few changes in words of symbolic importance, this statement and that of BSSRS are nearly interchangeable.

The fun began quite soon. In the fifth issue (8) was a long statement by one of the magazine's founders, Peter Harper. His opening set a style for much that was to follow:

I can hardly bring myself to say it. It feels like kicking my grandmother. But... pass the word friends... I think... A.T. IS DEAD. It was a wonderful dream, and we, the children of Hamlin, followed it dancing through the streets. I'm glad we did. On the walls of Paris in '68 they wrote 'The revolution which is beginning will call into question not only capitalist society but industrial society... We are inventing a new and original world. Imagination is seizing power.' *Imagination is seizing power.* I always find such romantic bullshit irresistible, as do many of us. Combine that weakness with our faith in that, for all our protestations, we all believe to be the real god of the Universe — science — and you have an idea whose appeal is irresistible. Well, it's led me a merry dance for three years, and now it's time to move on.

The main burden of his argument was that the old 'unified' conception of Alternative Technology (A.T.) was dead. The different criteria of quality (ecological harmony, personal fulfilment, small-scale, etc.) were not necessarily compatible on any particular project. He warned his friends that "we rejoin the human race, and with it the unavoidable conflicts and trade-offs of economics and technology".

Thus, at a stroke, all the inconsistencies and contradictions of the 'Alternative Technology' movement were laid bare. What was the response of Peter Harper's colleagues, indeed of himself? They did not withdraw for theoretical clarification before proceeding; nor did they redouble their efforts as their goal disappeared. They did something very new in politics. They carried on, joking all the way. Their fifth anniversary issue (9) has a review article with the title 'Still Crazy After All These Years'. Readers are treated to a glimpse of a policy discussion under the mocking headline 'The Party Line'. It is introduced by a sample:

'Why don't you build a windmill?' we shout at each other. 'Explore inner space!' 'Stop ignoring the occult!' 'Tap the energy of ley-lines!' 'Live in a commune!' 'Distribute your income equally!' 'Be a socialist!' 'Don't be a socialist!' And so on. Reading reports of

Undercurrents meetings, I get a picture of half a dozen people trying to give me directions. Some of them are telling me to turn left at the crossroads, some. . . to turn right (or) to go straight ahead, (or) to turn round and go back the way I've been. But the crazy thing is, non of them has found out where it is I want to go!'

For those whose conception of knowledge and action derives from the European rationalist style, it must seem a miracle or a mistake, that in such total absence of clear and distinct ideas anything regular and worthwhile could be accomplished. It is neither; rather, a supple and self-aware handling of the contradictions in basic beliefs among the editors, enables a consensus on content and style to operate. There seems to be a core of beliefs and commitments, positive and negative: that 'the system' has all sorts of things wrong with it, including oppression of the weak (of our species and others) by the strong: the truncation and distortion of awareness; and the industrial rape of the earth. But existing socio-political solutions are seen to be largely a part of the problem; and so the most practical thing to do, for now, is to draw on the Utopian radical political traditions, on small-scale technologies of a Schumacherite sort, and on the expanded consciousness discovered by the counter-culture of the '60's. There is no synthesis of all these elements; but then there is no genuine intellectual synthesis of any other worthwhile set; hence they carry on laughing. One explicit response to the ideological crisis in 'Alternative Technology' and the co-optation of the title, has been to shift to 'Radical Technology'; it has a better flavour.

The contents of an issue of *Undercurrents* will be a mixture of articles on 'hardware' (wood-burning stoves, windmills, amateur radio, energy systems), 'radical technology' (the Lucas initiative, educational and co-operative developments). 'critical studies' (of environmentally polluting or politically abhorrent technical systems) and 'consciousness' (D.I.Y. marijuana and mushrooms, Findhorn, ley-lines). The overtly political commitment is strongly maintained in the short articles of the 'eddies' section; there, the anti-State technology and struggles over the Official Secrets Act have prominence.

Every issue contains several pages of letters, frequently including some from disgusted subscribers who have discovered suddenly that *their* idea of A.T. has been left behind by the journal. They are a constant reminder that there is as yet no synthesis. But the journal goes on, its editorial gang appearing to enjoy themselves, and showing fewer worries abour finance, circulation and volunteer help than *Science for People* or many another journal on the radical fringe.

Such security, however firmly based on insecurity that it may be, could be a warning that the tendency served by *Undercurrents* has found itself too easy a path. It may be more fundamental and in the short run it is more rewarding, so cultivate one's own garden rather than to do pastoral agitation among the urban oppressed classes. Certainly, among the founders of A.T./R.T. there is no lack of awareness of the old-fashioned sorts of exploitation, and no lack of sympathy with its victims. But they may find it difficult to be involved in traditional radical politics because of their awareness that revolutions have had a tendency mainly to turn oppressive systems upside down. Hence their communication with *Science for People* and (more important) its intended audience, tends to become strained and unsatisfying; and they inevitably tend to become encapsulated in their own communal world. However, if there were ever a radical movement designed to slip through contradictions, the *Undercurrents* group is one.

5. Some Related Journals — A Note for Perspective

If the survival of our high-technology civilisation were not an open question, then small fringe journals like the two described above would be of only minor interest. But no-one now dares predict an easy passage to the end of this millenium; and in times of instablility a radical's fantasy can become (perhaps only briefly) State power. Hence, the 'anti-Establishment science' movements should be seen in the context of the crisis of confidence of our civilisation as a whole. At this level, the analogies between modern science and traditional religion become relevant.

Two journals have a particularly close relation with *Undercurrents*; the differences between the formal connections in the two cases is instructive. The journal *Resurgence* was founded by survivors of the radical pacifist groups that received great stimulus from the movement for 'nuclear disarmanent' of the late 1950s and early 1960s. It has been closely associated with E.F. Schumacher and has published many of his essays including some not reprinted elsewhere. Its main explicit message is for small-scale communities, cultural diversity, and self-sufficiency. It shows a deeper sympathy with the world of inner experience than does *Undercurrents*, being informed by meditational awareness rather than occasionally experimenting with earth-magic. There will be occasional articles of a technical nature, and the two journals even tried a joint issue once. But *Resurgence* has no interest in the ordinary sorts of politics, and the unspoken assumption that the world could so easily be a nice place (reminiscent of the Quakers) will repel those who have survived in politics only by foresaking that particular comfort.

Closer to *Undercurrents* technically, but very separate politically, is *The Ecologist*. This was the creation of Teddy Goldsmith, one of the most strident gloom-and-doom popular ecologists of the later 1960s. In one respect, the approach of *The Ecologist* is rigorously non-political. It believes that the coming ecological catastrophe will engulf us all, and it apparently finds no particular section of society ready to act as an instrument of salvation. Hence its criticisms are of the system as a whole, and its message is for a total reconstruction. Since this would be along anti-industrial lines, its technical material overlaps with that of *Resurgence* and *Undercurrents*.

Also, some political campaigns with a communitarian or libertarian ideology are given sympathetic reporting in *The Ecologist*, when they are focussed on nuclear power or other manifestations of high technology. But in Goldsmith's own vision, stability (ecological and social) is valued above all else, and he is led to attitudes on social organisation and the means of social change that the political radicals of any complexion find repugnant. Hence, his journal maintains a silent co-existence with the others.

In one sense, these assorted small-scale journals with their editors and audiences, can be seen as attempts to put together some of the fragments created by the explosions of consciousness of the 1960s. For the mainstream of our political and intellectual culture, and even for the traditional Left adhered to by BSSRS, that is all over and a part of history. Students are back at work, dope is domesticated, and our most pressing social problems are more reminiscent of the 1930s than of the 1960s. Scientists are generally supposed to be kept on a shorter lead than in the post-war period, and 'relevance' is an explicit criterion of choice in the determination of research strategy. But for most individuals and institutions, the game is played much as ever.

However, the world as seen outside the academies is irreversibly, and in some ways radically, changed. It is not merely that the 'environment', with pollution and resources problems, is a permanent part of the scene. Now we all know as citizens (though we rarely learn it as students) that uncontrolled industrial change may be bad rather than good, and that political decisions must govern technological choices. Also, there is the world of 'para-science' (which will simply not go away when subjected to an occasional denunciation by established scientists) to remind us that the cosmological questions were not finally settled by the generation of Galileo and Descartes.

It is impossible to predict the future of these continuing ferments on the fringes of the scientific culture. Even if they are destined to produce a transformation of our intellectual world, that may yet be generations in coming. But the limits of the analogy with religion that I mentioned at the beginning of this survey may provide us with an important clue on where *not* to look for evidence of the change. The fact that established science is *not* a total institution like the earlier established Church means that critics and visionaries can develop their work in isolation from it. They do not need to debate with University teachers and researchers; they can hope that some will come to adopt their way, and the others drift into irrelevance.

The abandoned church buildings of our post-Christian society are not simple and direct results of the debates on Evolution, and the victory of Science over Religion. Rather, they are a reminder of how very vulnerable social institutions can be to only moderate changes in the functions required by their environing society. Without being able in any way to predict the sort of transformations or convulsions that we may be approaching, and still less the shape of their outcome, I can still feel that these journals and their supporting groups have something to tell us about the sort of total response that would be appropriate to a time of crisis, and where established science shows little sign of ever having anything to say.

Notes and References

1. *Science for People* (Women's Collective Issue) No. 29, Spring, 1975.
2. *Science for People (Health Group Collective Issue)* No. 38, Winter, 1977/78.
3. *Science for People*, No. 37, Autumn, 1977.
4. *Science for People*, No. 35, Spring, 1977.
5. Hilary Rose and Steven Rose, (Eds.) *The Political Economy of Science*, Macmillan, 1976, and *The Radicalisation of Science*, Macmillan, 1976.
6. Young, Robert, 'Science *is* Social Relations', *Radical Science Journal*, No. 5, 1977, pp. 65–131.

7. Goodman, John, *Radical Science Journal* No. 4, 1976, pp. 8–9.
8. Harper, Peter, *Undercurrents* No. 5, Winter, 1973–74.
9. Editorial, *Undercurrents* No. 20, Spring, 1977.

The journal described above are not of the sort routinely purchased by academic libraries, hence I list below their addresses and subscription prices at the time of writing.

Science for People is published quarterly by BSSRS, at 9 Poland Street, London, W.1. The price for non-members in U.K. is £2 for individuals, £5 for institutions.

Undercurrents is published bi-monthly at £3.00 U.K. (6 dollars U.S.) from 12 South Street, Uley, Dursley, Gloucestershire, England.

Resurgence is published bi-monthly at £3.50 (7 dollars U.S.) for individuals, £5 (10 dollars U.S.) for institutions, form: Pentre Ifan, Felindre Farchog, Crymych, Dyfed, Wales.

The New Ecologist, combined with The Ecologist Quarterly, cost £7 (U.S. 14 dollars) from 'Ecologist', 73 Molesworth Street, Wadebridge, Cornwall, PL27 7DS, U.K.

KNOWLEDGE AND OPINIONS

ROBERT FRANCK
Catholic University of Louvain

Introduction
I.   *A Way of Thinking*
   1. Thought is cut into two.
   2. It is unquestioned ...
   3. ... and it is very useful.
II. *The Ideology of Knowledge*
   4. The strange amalgam of dominant thinking.
   5. Objectivity is ideological.
   6. The ideological role of objective knowledge, when it is opposed to subjective opinions.
III. *Ideology Linked to the Natural Sciences.*
   7. Ideology based on the natural sciences.
   8. Different perspectives.
   9. The sciences are stripped of their critical power.

Introduction

It is customary to oppose knowledge to opinions. But what constitutes knowledge, and what is only opinion? What are the criteria for making such a division? The fact is that today we have at our disposal many criteria for making the division, and it is not difficult to enumerate them. To quote in random order, by way of example: observation, reliable evidence, rigorous application of method, the use of a precise language testing by experiment, power of anticipation, formalisation, ability to proceed to applications, usefulness of these applications, accountability to certain principles or to acquired knowledge, impartiality, accepted views of specialists, etc. Usually we bear in mind a few of these criteria and attempt to coordinate them; whereas others, which we do not consider indispensable or relevant, are put aside; *this varies* according to the knowledge envisaged ... And one would be in greater difficulty trying to find a justification for such criteria! But this is not our purpose here. If we mention it, it is only in order to make it clear that we will not be concerned with such matters, nor with such question as when it is permissible to characterise this or that as 'opinion' or 'knowledge'? Our subject is different: we want to raise questions concerning the use which is made of this opposition between knowledge and opinions, concerning the ideology which underlies this opposition, and concerning the resultant effects for the sciences.

In what manner does one most frequently oppose knowledge to opinions? We incorporate knowledge into a certain prevailing conception of science which is characterized as objective. To this is opposed all the rest, which is grouped haphazardly under the names of *opinions* and *value judgements*; these are characterized as subjective. In this way we provide a justification for the division we make, for the wall we erect between scientific practice and everything which is said, written or thought outside of the established sciences; and we prevent the sciences from being critical. One can say more: opposition between knowledge and opinion, in its present form, partially determines the actual status of the sciences, and wrongly assigns limits to them.

The opposition between knowledge and opinion did not arise from the sciences; it comes from the dominant ideology. It is this opposition which determines to a certain degree the present day status of the sciences, and wrongly assigns them their limitations.

The division thus created between science and 'objectivity' on the one hand, and opinions and 'subjective' values on the other, is accompanied by a disqualification of what does not deserve the name of 'science', a disqualification of 'subjectivity' and 'value judgements'. There is also a disqualification of the sentiments, and needs, and the desires of the individual, of his or her life and death, in the face of the 'objective' demands of 'reality': this permits the justification of 'realistic' policies of oppression and repression.

The real alternatives confronting us are the following: between the interests of an individual and those of the larger social group to which he belongs; between the interests of different groups and classes; between the interests of the masses and those of people in positions of authority. These alternatives are deformed and retranslated in terms of subjectivity (it will be said that complaints are subjective) and objectivity (it will be said that decisions made by authorities are determined by objective realities).

The opposition of objective knowledge and subjective opinions allows demarcation lines to be drawn between what is acceptable and what is not to those in power. Finally it is the 'knowledge' established and institutionalized by the ruling class which is used as objective knowledge. Any other knowledge is disqualified.

But in reality, the opposition between knowledge and opinion is not unquestionable, it has its rationale in the dominant ideology, and rests upon a conceptual amalgam which can be undone.

These are the principal ideas which will be developed in the following pages.

I. A Way of Thinking

1. *Thought is Cut into Two*

By opposing, as one most frequently does, 'objective' knowledge and 'subjective' opinions, one consolidates division of thought into two autonomous fields, each isolated and foreign to the other. On the one hand, the field of opinions, or of the subject, or of what one wants, where at leisure and without harm — because this is outside reality — man can nourish dreams or aspirations, pursue ideals and cut down ideas. On the other hand, the field of knowledge, of the object, or of what is: here one is realistic and one remembers that things, as they say, 'are what they are' (1). What happens to truth in all this? It is assigned a place beside the object. As for values, they are left to the subject; he will dispose of them as he sees fit.

2. *It is Unquestioned...*

This is not to claim that this conception of the two fields of thought, each completely isolated and hermetically sealed from the other, with truth on one side and values on the other, is the only one in current use. There are other ways of conceiving the practice of thought. On the other hand, looking more closely, one sees that day to day language, just as scientific research, does not follow, as much as one might think, this dichotomy between truth and values (2). Nevertheless the concept of two independent fields of thought (values and truth, opinions and knowledge) is most commonly accepted by everybody — ourselves included — and it dominates social life.

It is for this reason that, as often as not, in the field of opinions it seems to be impossible to decide between those who are wrong and those who are right; it is common sense that we would not let ourselves be convinced by those who do not share our ideas, just as we would not be able to convince them either; it is common sense that 'culture' is practiced beside or above 'real' life in a closed arena; it is thought right to abandon opinions to free competition ... All this shows that opinions are not really believed to have much connection with truth; rather they depend on sentiment, passion, free choice or interest.

There are a thousand opinions on life in town or country, on traffic, unemployment, education, the Olympic Games, the Police, and everyone makes his own choice. The opinions people have may be respectable, but they are not very serious ...

3. ... and It is Very Useful

The importance of the distinction between 'objective' knowledge and 'subjective' opinions appears in a flagrant and brutal manner in the use currently made of it in politics. This opposition must be placed in a larger framework, namely that of the semantic opposition between subject and object, which is a determining factor in many present-day modes of expression. Shortly, we will see the ideological scope and ambivalence of the opposition between subject and object.

In this opposition between, on the one hand, the subject, bearer of needs and subjective aspirations which are translated into opinions which are neither true nor false, and on the other hand, social, economic or military reality, which alone is objective and true, it is only right, if one of the two must predominate, that it should be reality. One must be 'realistic'. It becomes reasonable and normal to impede the expression of individual sentiments which threatens established order. There is nothing wrong in manipulating people's opinions and needs through propaganda, advertising, and electoral campaigns, in exploiting and even killing people, if (and only if . . .) reality demands it. A clear political conscience finds in this it most recurrent and banal argument. Oppression and repression at home or abroad — prisons, dismissals, wars — are justified by highly 'realistic' arguments such as social stability, economic expansion, the balance of forces, and so on. The life and death of individual human beings is not part of reality, they are placed together with the subject, subjectivity, opinion, and values. Death, mourning, imprisonment, deportations, unemployment, grief, separation, brutishness, humiliation, are all matters of sentiment; sentiment must not predominate over reason, or over the 'objective' demands of reality.

This does not mean that the dualism — subject (subjectivity)/object (objectivity) — which creates the alternative between objective knowledge and subjective opinion, is the *cause* of a policy of oppression and repression: it is simply the outward form, the 'grid' through which political reality can be read. At the same time the subject/object dualism gives to this policy its *a priori* justification and appearance of rationality. (3)

We can find an extreme (satirical?) illustration of this concept in a small book entitled *Report from the Iron Mountain*, foreword by H. Mc. Landress (J. K. Galbraith) 1968. This would appear to be a report written in 1966, after two and a half years' work, by a Special Study Group, and intended for the American government.

To give an example, in its Recommendations to the government, the report stresses the need to study

how to compute on a short-term basis, the nature and extent of the loss of life and other resources which should be suffered and/or inflicted during any single outbreak of hostilities to achieve a desired degree of internal political authority and social allegiance (4); how to project, over extended periods, the nature and quality of overt warfare which must be planned and budgeted to achieve a desired degree of contextual stability for the same purpose; factors to be determined must include frequency of occurrence, length of phase, intensity of physical destruction, extensiveness of geographical involvement, and optimum mean loss of life; (p. 198).

The approach adopted by the Special Study Group is clearly defined;

what they wanted from us was a different kind of *thinking*. It was a matter of approach. Herman Kahn calls it "Byzantine" — no agonizing over cultural and religious values. No moral posturing. It's the kind of thinking that Rand and the Hudson Institute and I.D.A. brought into *war* planning.

Better still:

There *is* such a thing as objectivity, and I think we had it. . . . I don't say no one had any emotional reaction to what we were doing. We all did, to some extent. As a matter of fact, two members had heart attacks after we were finished, and I'll be the first to admit it probably wasn't a coincidence. (p. 35).

Here is a second example, taken from an article by John D. Williams, of the

Rand Corporation: "The absurdity of safe driving", *Fortune*, September, 1958.

> I am sure that there is, in effect, a desirable level of antomobile accidents — desirable, that is, from a broad point of view, in the sense that it is a necessary concomitant of things of greater value to society.

II. The Ideology of Knowledge

4. *The Strange Amalgam of Dominant Thinking*

The subject/object dualism, along with its associated connotation of subjectivity/objectivity, also has another connotation, which is linked to rationalism. If we say *this* rather than *that*, this is perhaps because we are under the influence of our subjectivity, or perhaps because the objectivity of facts forces us to do so. But it can also be by reason of what has been said before, or of what could be said: we then say that we are guided by 'reason' itself (neither subjective nor objective, but rational . . .). But if reason alone is to be heard when we speak, then what we want, our needs and inclinations, must be neutralised. Thus every man is led (sooner or later), if he submits to the laws of reason, to think the same thing. Reason is autonomous, it is subordinate to neither the subject nor the object, it has its own justification and it determines its own laws. This produces a new dualism. The subjectivity of the speaker is eclipsed; speakers become interchangeable and become nothing more than the vehicle of reason, and in front of reason or the rational subject we find 'data' which lends itself to a rationalisation (4).

This conception, which can be crudely characterized as 'rationalist', has a long history, which goes from Galileo to Einstein and Planck. It is not however this history which concerns us here, but what could be called the common sense rationalism which confronts us today; how can this rationalism coexist with the subjectivity/objectivity dualism?

In fact the two dualisms, that of subject (subjectivity)/object (objectivity) and that of subject (reason/object (data), do not at all correspond. On the contrary, reason will take its place at the side of objectivity and is presented as an alternative to subjectivity. Reason is opposed to subjectivity. Reason, as everyone knows, 'orders' and 'controls' Nature, in the same way as it controls feelings and passions. The result is that when the two dualisms are amalgamated in the way this is done today, reason and objectivity become ranged on one side and Nature and subjectivity on the other.

At the same time the opposite can be said: it can be asserted that Nature is objective reality and that reason must be placed at the side of the subject. Once the two dualisms have been superimposed, almost anything can be said. For example: subjectivity is, because of emotions and passions, the *natural* element which remains in the subject; but at the same time it is the most *personal* dimension of the subject ... The reason (!) for this is that we are dealing with two heterogeneous modes of thought. They have a different historical origin, and are not made to go together. The reason-Nature dualism developed in what Michel Foucault called the *epistēmē* of the 17th and 18th centuries, and the subjectivity-objectivity dualism in what he calls the *epistēmē* of the 19th and 20th centuries (5). But although the two modes are not made to go together, in practice they are indeed used together (they are both very much alive in their daily usage). Why?

When these two modes of thought are used together, *almost* anything can be said, but not *absolutely* anything! There are two things at least which are established as constant and which it is impermissible to deny: objectivity has, of right, the upper hand over subjectivity and reason has, of right, the upper hand over Nature. Subjectivity and Nature are inferior. We (subjective) subjects have to submit to objective reality, we (rational) subjects have to control Nature, bring it into submission. These two imperatives formulated by the two heterogeneous modes of thought, are of opposite meaning; it would be very difficult for us to reconcile them in practice if social organisation did not come to our aid; the subjects who submit and the subjects who bring into submission are not the same! Either this is only a semantic game, or it is the beginning of an explanation for the aberrant and unnoticed existence of two heterogeneous modes of thought. This amalgam may be explained by the role it plays in justifying and more generally maintaining social structures.

We can now deal with three interlinked questions:

(a) The ideological role of the subject-object dualism

(b) The ideological role of the hybrid superimposition of the oppositions of subjectivity/objectivity and Reason/Nature

(c) The ideological role of the opposition between knowledge and opinions.

5. Objectivity is Ideological

Louis Althusser says that ideology designates to individuals the place which must be theirs (6). For example, I am 'a worker, a boss, a soldier' or a professor; but to take another example of Althusser's, I am also, by virtue of the ideological structure of the family, a boy or a girl (p. 32). Finally I can, through the religious ideology of Christianity, see myself as a subject *of* God, a subject subjected to God, subject through the Subject and subjected to the Subject (p. 34).

From putting forward this analysis of the religious ideology of Christianity, Althusser deduces — perhaps too hastily — that *every* ideology calls upon individuals as subjects in the name of a single and Absolute Subject. Every ideology, he writes, is *centred*; "the Absolute Subject occupies the single place at the centre, and calls upon around him the infinity of individuals as subjects, in a two-way specular relation of such a kind that it subjects the subjects to the Subject, while in the Subject where every subject can contemplate his own image (present and future), it gives them the guarantee that it really is a question of them and of him (. . .)" (p. 35).

This theoretical definition of all ideology, which is proposed by Althusser, helps us understand the ideological role played by the dualism of the subject and the object. In this dualism *the subject is placed opposite the object*, and no longer opposite the absolute Subject (God), described by Althusser for the religious ideology of Christianity. This represents a great upset. And one can see in the history of the European bourgeoisie the signs of the struggle for the triumph of the subject/object ideology over another ideology, that of subject/Absolute Subject (God). The revolutionary bourgeoisie of the 18th century refused to accept an ideology of subjects 'centred' on the absolute Subject, depositary of authority and centralised power, and accepted no constraints except those which came from the order of things and from Nature, or again from Reason, and it was this order and this Reason which in their turn defined subjects as having equal rights (natural rights, the Rights of Man). But the bourgeoisie, when it came to power, dressed its needs with another sauce. It is here that we see the emergence of the amalgam of the subjectivity/objectivity and Reason/Nature oppositions. The struggle continues under the same flag (the flag of Reason and the observation of Nature); it can also continue with the same intentions among its principal leaders, but it is diverted to the service of another cause: it is no longer a question of contesting the old authority of a centralised and absolute power 'subjecting' 'subjects' through the grace of God, but of strengthening bourgeois power. In other words, subjects must not, in respect to the 'place' they have been assigned opposite the object (Reason and Nature), really have equal rights, but they must think they have. (Otherwise bourgeois ideology would turn against bourgeois power.) How can the conjunction of the oppositions of subjectivity/objectivity and Reason/Nature achieve this objective? It fulfils its ideological role by parasitically attaching itself to the bourgeois revolutionary ideology of subject/object, and drawing off all its strength and power of persuasion.

To follow this, it is necessary to recall the two opposing imperatives which flow from the subjectivity/objectivity and Reason/Nature amalgam: we (subjective) subjects must submit to objective reality, we (rational) subjects must control Nature and keep it under control. Under these conditions what happens to the 'place' assigned to the 'subject' opposite the 'object'? The subject is at the same time *above* the object and *below* the object! (The subject dominates the object and 'subjects' itself to it.) It is not only a very uncomfortable position for anyone without the gift of ubiquity it is also contradictory; unless that is, those who bring into submission and those who submit are not the same. But who is going to bring into submission and who is going to submit? One can say that in relation to the object (with Reason, Nature, and the nature of things) *everybody* is a subject; and in turn everybody is called upon both to bring into submission and to submit. The bourgeoisie no longer knows, or pretends no longer to know, who is the master and who is the slave. For everybody is *called upon* to bring into submission, i.e. to assume responsibilities and powers according to his merits and competence, and if everybody is in turn called upon to submit, it is no longer to masters that he has to do so, but to law. Masters no longer exist, there are only leaders, themselves only the guarantors, the preservers, the guardians – in short the servants – of this order. These servant-leaders are themselves *subjected* to the order and the laws of society which are no more personified than the laws of Nature; they are subjected to them just like 'anyone else'. (And woe betide that leader who weakens belief in this ideological schema by some tax-evasion or abuse of language.)

Unlike religious ideology which explicitly consecrates the authority of those who have power by basing it on the absolute Subject, here authority bases itself on the nature of things, on Reason and on objectivity, which also have absolute value. Thus, it is now important to characterise contemporary politics as 'rational', 'necessary', 'realistic', 'objective', and even 'scientific'! In short, those who exercise authority justify it by claiming that they are on the side of reality, of objectivity, and of reason; as for those they keep under control, they are on the side of subjectivity (ignorant) and of Nature (uncultured) (7). But at the same time we are taught that *all subjects are, at the same time*, subject to reason, that all have both to bring into submission and to submit, and that all are given the same culture, the same Reason, the same knowledge.

One might say that the subject/object ideology, as transformed by the subjectivity/objectivity and Reason/Nature amalgam has become a *substitute* for the subject/Absolute Subject ideology. Here the Absolute Subject is no longer God-king but 'Reason' and 'objectivity'; which now 'occupy the single place at the Centre', which is also the place taken by those who have power; and it is they which class the infinity of individuals as subjects. Let us paraphrase a formula of Althusser's concerning religious ideology (p. 35). It is in reason and objectivity (and sometimes more spectacularly in the show and glory of "scientific" reason and objectivity and their technological display) that the subjected subject now contemplates its own image (the image of a subject equal to all subjects, still tainted with subjectivity and non-culture (Nature), but called to Reason and universal objective knowledge) and finds the guarantee that recognising itself in this image it will be saved . . .

Is it excessive here to speak of *salvation*? Here is a learned example of this salvationism:

Perhaps even more than an 'explanation' (. . .) man needs to go beyond himself, to transcend himself (. . .) No value system can pretend to constitute a real ethic, unless it proposes an ideal which transcends the individual to the point of justifying, if need be, his sacrifice. By the very loftiness of its ambition, the ethics of knowledge could perhaps satisfy this demand to go beyond. It defines a transcendental value, true knowledge, and proposes not that man should make use of it, but that from now on he should serve it through a deliberate and conscious choice'. Jacques Monod, op. cit. (p. 192).

Faith in 'progress' through science is another version of this salvation, less 'noble' and more current than that of Monod.

6. The Ideological Role of Objective Knowledge, When It is Opposed to Subjective Opinions

If knowledge and opinions are opposed to each other, it is due to the ideology we have just outlined. The role of this opposition is to draw different lines of demarcation between what is acceptable and what is not acceptable for those who are in power.

Objective knowledge, *as opposed* to subjective opinions, *determines* 'objective reality' in relation to which the 'subjects' must define themselves and submit. What we learn from objective knowledge must not be put in doubt; it would be sheer madness not to submit to it . . . And it would be sheer madness not to submit to the decisions of the powers that be when such decisions are the fruits of this knowledge. (8)

Objective knowledge is universal, it is the same for every subject; and all the better if, as though by chance, the imperatives of knowledge regularly match the objectives of the authorities (at all levels and in all fields). Let us see how this happy chance occurs . . .

What are the imperatives of knowledge which decide what is acceptable and what is not . . .? They come straight from the scientific ideal. They are observation, faithfulness to facts, the obligation to keep to reality, operative value, and scientificity as the exclusive access to the truth. In reality these imperatives of knowledge have an imprecise content, they are rather like invocations, and all send us back to the even more vague requirement of 'objectivity'.

(a) *One must keep to what is observable*

What is not observable is not acceptable. It follows that the feelings and desires of people cannot be taken into account in the decisions which must be taken for security, defence, economic growth, order, legality, profitability, educational selection. (As has been pointed out.) This imperative consigns to subjective non-reality not only 'inner' life (feelings, desires, needs), but also everything connected with individual interests (including life and death).

(b) *One must keep to the facts*

This means that 'value judgements' which call into question the facts endorsed by the powers that be can be ignored. These value judgements are rejected as "subjective opinions", or as partisan options, stripped of objectivity.

(c) *One must keep to reality*

To promote changes not foreseen by the authorities and which are not already sketched out in established 'reality', becomes an unrealistic enterprise. It follows that to forbid these changes is to make a show of reason.

(d) *One must keep to what is efficient*

This imperative is in competition with the contemplative ideal of science. But it is powerful. The idea of Bacon's, of science as a producer of wealth, has become the powerful ally of industry. Science is steadily turning into technology and from now on wants to be 'operative'. It is not by chance that the word 'operative' is (improperly) used to refer to efficiency. One must keep to what is efficient, so that any reality which is not in accord, either directly or indirectly, with the aims of production, or which does not appear to be profitable, is consigned to non-reality.

(e) *One must keep to what is scientific*

Anything which does not belong to 'scientific' knowledge is unacceptable. Outside of science there are only idle chat and subjective opinions. But what is scientific and what is not? The words 'science' and 'scientific' are used many ways; today they are used to designate the so-called Natural Sciences, as well as the so-called Human Sciences; even sexology, pedagogy, journalism and philosophy are now 'scientific disciplines'. If we take a closer look, the label 'scientific' simply covers all university and academic 'knowledge', the 'knowledge' established and institutionalised by the class in power. Anything which is said outside of this established knowledge is not recognized as scientific.

III. Ideology Linked to the Natural Sciences

7. *Ideology Based on the Natural Sciences*

The ideology which opposes 'subject' and 'object', with its connotations of objectivity, Reason, subjectivity, Nature, is closely linked to the 'Natural Sciences'. What is this link?

A scientific practice is always linked to a *general* discourse which goes beyond particular practices: for example, one which asserts the demand of objectivity, or one which puts forward the operative ideal. Such a general thesis gathers together the different scientific practices: it is this which serves as a common denominator, which makes them a part of 'science' in spite of their diversity. This general discourse is not simply superimposed onto scientific practices from the outside, in the same way as a 'philosophical' reflection on the sciences might be; it intervenes in the determination of methodological requirements and it partly determines the meaning, the orientation and the limits of these practices.

But at the same time, the general discourse seems to lead a double life. On the one hand, it forms a part of sciences, to such a degree that it is through it that a certain number of practices are included in a group which can be identified as being the family of 'sciences'; on the other hand, it joins in many other practices which are quite different from those called 'scientific'. We call in scientific objectivity at every turn and at every event, as we said at once, to draw dividing lines we are wanting. With regard to operational ideal which is gaining ground, we refer to it for example about business management or about military strategy or pedagogy ... (9).

We could suppose at first sight, that that 'double life' to which the general discourse of science seems to lead, only results in the fact that we indulge too freely in it — it would be sufficient not invocating the principles and general contraints of science into areas which are unknown to it. But one can also propose that if these general contraints are so imperative inside as outside sciences, it is because they are coming from prevailing ideology. Otherwise, it is because scientific practices are defined across prevailing ideology of knowledge, it is because they are looking for their oneness or their common denominator into that ideology, it is because they accept that ideology as a general discourse of 'the' science, that that (ideologic) discourse also impose itself as inside as outside sciences.

We may say now that it is 'the' science (it is a group of united practices owing to a same general discourse) which constitute the basis of the knowledge ideology. For it is about science that the opposition between subjectivity and objectivity, which connotes and determines the ideology subject/object, gets its guarantee. Science by its very existence and by its success gives substance to the exigency of objectivity against subjectivity and the credit of which disposes ideology coincides with the credit which possesses science.

All this does not imply that the sciences are by 'nature' in solidarity with this ideology: there is nothing to indicate that the general thesis of the sciences cannot change and redefine the scientific practices in terms that are foreign to this ideology, a redefinition which would be accompanied perhaps by modifications in this practice. In relation to this let us consider two contemporary perspectives in the sciences.

8. Different Perspectives

There is first of all, as we quoted above, the operative ideal which tends to substitute itself today for the ideal of objective observation, and which is alien to the oppositions of (subjective) subject/(objective) object and Reason/ Nature. Does science deprive so much prevailing ideology of knowledge from its guarantee? Surely not. Because the way by which one commonly opposes knowledge and non-knowledge is not put in doubt by the operative ideal. This rather reinforces the prestige of scientific knowledge, makes even more ludicrous everything that is not connected with 'science' and by emphasising the opposition between knowledge and non-knowledge, lends its support to the objectivity/subjectivity opposition. So much is this the case that there is frequently confusion between the ideal of objectivity and the operative ideal; thus for example the genuine(!) 'objectivity' of any scientific matter is judged by its 'operative' character!

The second perspective is what certain Anglo-Saxon researchers have called 'critical science'; it consists in wanting to relocate scientific practice in its socio-political and cultural background. It is a question of confronting 'science' with that which is not science, of calling it into question according to non-scientific criteria; in short, of destroying the wall which exists between 'knowledge' and non-knowledge'. Such an approach, undertaken by scientists, can lead to the dominant ideology being deprived of the support given by the sciences. For it asserts that these scientists no longer *believe* that there exists a *region* of objectivity ('science') where the whole truth can be found, a gilded temple of knowledge surrounded by the darkness of ignorance and subjectivity, with some people inside and some people outside. These researchers, on the contrary, witness the necessity of judging the 'inside' from the 'outside' . . .

This approach has the value of a testimony. But in addition to this testimony this 'critical' approach to the sciences has the advantage of showing that the sciences are *not only* a language and a unity of experimental practices, but that they are also a social, political, cultural and economic practice.

A socio-political analysis, although in juxtaposition to scientific theory, does in fact allow us to consider the connections between scientific theory and its conditions and finalities; not of course from the interior of scientific theory, as though one could open up a football and find inside the passing match. It is too easy to maintain that the connections between 'science' and 'the rest' are difficult to conceptualise since these connections are not visible, and cannot be formulated in the interior of scientific theory. It is true that scientific theory possesses a relative autonomy: it must not be confused with the conditions of its production, it does not change according to the weather, it can be the same in New York and in Peking. But the question is to reverse the perspective and instead of looking inside scientific theory for the traces of what is external to it, instead of being astonished at not finding the whole in the part, it is a matter of locating the (historical and changing) role of scientific theory and scientific practice, in socio-political life.

Moreover, if the socio-political analyses of scientific practice are 'exterior' to it, if they are *juxtaposed* to the exercise of the sciences, they nevertheless show the (political and scientific) necessity of a further undertaking: that of establishing a new general thesis (discourse) of the sciences.

The present general thesis of 'science' excludes from consideration the extra-scientific conditions and objectives of scientific practice (10). This thesis demands, in the name of scientific objectivity, that science should obey only its own imperatives, that its development should be ordered by nothing but its own *internal* logic. The general thesis of 'science' masks the relationship that in fact exists between the specific requirements of the sciences and social requirements, and it masks the fact that the perspectives of scientific research are subordinated to socio-economic constraints. The possibility can be considered that the sciences might be given a new general thesis. It would be a thesis which would define scientific truth in such a way that it was no longer juxtaposed to 'all the rest', and in such a way that scientific practice would again be understood as a social and political practice, with the choices which that implies.

If they were given such a thesis, the sciences would cease to be defined through the dominant ideology of knowledge, and they would deprive that ideology of the indispensable support that they give it at present.

9. *The Sciences are Stripped of Their Critical Power*

To the extent that the sciences are defined through the dominant ideology of knowledge, they are stripped of their critical power.

In the 17th and 18th centuries, the truth of the Natural Sciences had a real critical force which brought into the light of day the social and political scope of scientific practice. As we have already mentioned above, it was a question of opposing the order of Nature or Reason to the order of the Monarchy and the Church. To accept only the truth which comes from the nature of things is to reject the 'truths' which come from elsewhere or from above and which are imposed by *authority*. It is also to give *value* to reality as against dogmatism, beliefs, prejudices and the magic of words which conveyed the established values and justified the social order. We need not think that all the people concerned with science had such an objective, or that every scientific discovery contributed to it. It is sufficient to note that the scientific thesis was, by giving *value* to facts, in opposition both to the dominant system of thinking and to the ideology which subtended it (subject/Absolute Subject). The scientific thesis was playing a critical role, and this is confirmed by the various forms of opposition which it encountered.

But everything changes from the moment when the knowledge of facts, instead of calling into question what is said and what one tends (is obliged) to believe, claims that it is sufficient in itself. Facts for facts' sake. Objectivity for objectivity's sake. Criticism can only exist if the sciences are to be examined on what are today called 'opinions', which presupposes that they are to hear the questions asked of 'opinion' in the language of 'opinion'. Instead of this, the partitioning between the sciences and 'the rest' is total. *Truth*, instead of resulting from the confrontation of what is said and what is, becomes confused with facts and imprisoned in the closed arena of the scientific thesis. On the one hand there is the 'objective' scientific knowledge of the facts, and on the other 'subjective' opinions or value judgements. Thought is cut into two. And this is unquestioned only because of the prevalent belief that science is only made up of statements of fact obtained through neutral observation, through a pure contemplation of the Nature which surrounds us; also because of the conviction that value judgements are mere options which vary according to individual taste. Of course value judgements can be just that. Of course statements of fact exist. But to assert that the sun did not revolve around the earth, and that light was a wave, to carry out the dissection of human bodies, and to 'see' that the nerves originated in the brain and not in the heart — that, too, was to make a value *judgement* on established ideas, and moreover on the old way of establishing relations with Nature (it was to reject the dominant ideology of the time: subject/Absolute subject).

The subject/object ideology of the sciences in the 17th and, more openly, the 18th century, demanded that humanity should have no other master but the laws of Nature and of Reason. It rejected other masters. The ideology of the division between subjectivity and objectivity on the contrary obliges the sciences to keep silent, that is to say to ignore what is 'exterior' to them, and to close themselves inwards, to speak about themselves and to finish by saying nothing, to turn themselves into operative (or operational) machines seeking their place in the system of production. The ideal of the 17th and 18th centuries is still floating above scientific practice, but it is no longer linked to that practice; it is merely an academic and ornamental slogan which allows us to forget the real conditions under which the sciences are practiced today.

Notes and References

1. This way of opposing knowledge to opinions is illustrated in the well-known book of Jacques Monod, *Le Hasard et la Necessite, Essai sur la Philosophie Naturelle de la Biologie Moderne*, Le Seuil, Paris, 1970. It must be realised that on this point Monod has merely illustrated what is an everyday and generalised practice of the dominant thinking.
2. On this point, see the works of Mario Bunge, especially *Scientific Research*, Springer, N.Y. 1967.
3. While we must denounce the manner in which this permits the ruling authorities to justify the oppression and repression of people in the name of the so-called 'objective' demands of reality, it does not follow that we should only accept individual values as significant, and that social and economic realities are to be despised. In both cases the dualism of subjectivity/objectivity produces mystification, and one remains a prisoner of the individual/reality opposition.

4. The question is then asked, in terms of rational scientific knowledge, and first of all in terms of physics: What is ascribable to reason and what to data? Such a question is based on a presupposition: that of the dualism of reason and data. This dualism determines the very meaning of the question and anticipates the answer which will be given by allowing no way forward other than the division of the 'dowry' between data and reason. A dead end.
5. Foucault, M., *Les Mots et Les Choses* Gallimard, Paris, 1966.
6. Althusser, L., 'L'Ideologie et appareils ideologiques d'etat (Notes pour une recherche"), *La Pensee*, no. 151, June 1970, p. 29.
7. There can be no other culture, no other reason, no other knowledge, apart from those of the dominant class.
8. How can those who are in power make us believe that they are on the side of reality, objectivity and reason? By taking account of their knowledge. It is through the acquisition of knowledge (at school and at university) that one passes from subjectivity to objectivity. It is knowledge which guarantees the objectivity and rationality of the leaders.
9. It will simply be noted here that there is an ambiguity in the word operational which is connected as much to *operationalism* as to *operational* research ... Such an ambiguity is no accident. The two different interpretations of the word 'operational' translate the same option: that of putting in parentheses *reality*, or the question: What is it? What about the nature of things, their meaning and value? (What about light, matter ... , the good of man?) This is done to the benefit of another preoccupation which could be roughly formulated as follows: provided that it works ... (operational concepts, economy, functionalism-functional, economic, efficient ...) There is need for an analysis of how the ideological implications of that operational ideal are articulated in the ideology of objectivity we are analysing, but this is not our present purpose.
10. "... researchers who stood up against the arms race or against the war in Vietnam have, at the same time, continued to receive military funds. Everybody found that normal, because for researchers, research as such is the main priority, a priority which requires money (...) Science is considered as a neutral field with its own laws". *Les scientifiques et la course aux armements, entretien de Pierre Thuillier avec Milton Leitenberg*, in *La Recherche*, Paris, No. 19 January 1972, p. 16.

CAN THE UNITY OF SCIENCES BE CONSIDERED AS THE NORM OF SCIENCES?

AGNES HELLER
La Trobe University

In the second half of the 20th century, two programmes of the unification of have confronted each other. One of them represents an up-to-date version of positivist thought according to which unification ought to be realized as the universalization of the method of mathematized natural sciences. The other programme sets out from Husserl's *Krisis* and it advocates the thesis according to which the emergence of modern natural sciences is an historical achievement; consequently their world-constitution is reversible; the task of unification should be undertaken by a universal philosophy. I have to state at the very beginning: my sympathy lies with the critical partner (the Husserlian school), nevertheless I do not share its programme. It is this disagreement that I want to corroborate in the following arguments.

We are entitled to speak of authentic science whenever we meet the search for true knowledge and whenever this search is *de facto* separated from more traditionally inherited opinion, and, accordingly, not only in the instances in which this contradiction is itself transformed into the subject matter of knowledge, since this is already philosophical thinking. Because Husserl localized the source of all sciences in philosophy, it was consistent on his part to have transposed the date of birth of science into Greek antiquity. Undoubtedly it is the case that, specifically in Europe, science originated as philosophy. This specific characteristic, however, cannot be universalized.

Yet irrespective of whether the 'search for true knowledge' will take a specific shape in the form of philosophy or not, this specification occurs within the context of a concrete social integration. The contents and methods of the search for true knowledge, even its language (if it has a language of its own) will be led by the general value system of the integration whose science it is. Before the emergence of bourgeois society every integration had a fixed hierarchy of values. This hierarchy was equally valid and appropriate in all spheres of objectivation; from this point of view science, religion, philosophy and art were manifestations of a unified medium. Every integration had its *own* knowledge and within that framework its own science. Different types of mathematics, medical science, astronomy, 'co-existed' side by side; each of them fulfilled its function within the unified way of thinking and value hierarchy of its own integration and fulfilled it well. It has been brilliantly proved by Vernant (1) that the victory and prominent role of Euclidean geometry in the Greek city-state societies can be interrelated with the structure of the city-state itself. This differentiation of sciences according to integration has not entirely vanished even until now: I refer here to the separate Chinese medical science.

Where the totality of the knowledge of a given society is led by a unified hierarchy of values it is self-evident that no de-anthropologized world-view and attitude towards nature can come about. With the emergence of bourgeois society, however, societies as communities are dissolved (continuously and in the first instance in Western Europe) and together with this dissolution the fixed hierarchies of value disappear as well. The basically common language of *epistēmē* vanishes. There is no canon in art any more nor is the subject matter and function of portrayal determined by the collective (Christian) myth. Personal choice of value increasingly gains ground in the formulation of rival social theories; it is at this time that every historiographer begins to choose 'his' or 'her' own history.

This dissolution of the *sensus communis* creates a vacuum with respect to the attitude towards nature, and this vacuum is experienced as the legitimation of their own liberty and emancipation by the newly developing modern natural sciences. As a symbolic gesture, we generally link this turn with the name of Galileo. The natural scientist opposes, too, his or her own choice of value (according to the potentialities and demands of the given epoch) to traditional hierarchies of value: this choice is that of de-anthropomorphization as value. From this time on, nature is going to be considered as a mere object and manipulated accordingly. Thereby a new concept (and interpretation) of objectivity and truth comes about, and similarly the new language of natural sciences is given birth. This language is symbolic, it can be acquired on the basis of any given common language, tradition, social structure, stratum or class, it can be acquired equally and in an identical way. As a consequence, this language transcends all integrations and formulas and sets itself as being acceptable for all mankind. So it was the new (symbolic) language of natural sciences that *became* the *sole sensus communis* in an age of dissolution of integrations, communities, other types of *sensus communis*, the sole scientific language whose norm it is that it could be spoken by every one and in an equal manner.

This language has developed in an age in which the universal concept of humanity as abstracted from religion, race and nation was born. It is no surprise, then, that the new attitude had a powerful fascination. Philosophy was one of the first to be influenced by it and this is no surprise again since philosophy has always striven for the foundation of the universal validity of its ideas. *More geometrico* is the leading principle of the construction of metaphysics. Nor could historiography and social theory remain intact from this emanation. So history is — according to Möser — 'natural history' where everything could and should be explained by necessity.

Sceptical voices, however, appeared on the scene in parallel with this emphasis: there is a considerable price, they argued, attached to this new common language, to this new conception of objectivity and science. It has to pay the price of being abstracted from everything that is human, for the ever given societality, from value ideas of moral and non-moral type.

Kant is generally considered to be the thinker who provided a philosophical foundation for natural sciences. I think, however, that the thesis should be reconsidered and reformulated: Kant was the first to delineate the *limits* of natural sciences and to render this problem into the subject matter of his philosophy; while he acknowledged the legitimacy of the own language of natural sciences he placed human — moral — value *higher* on the scale than the concept of truth inherent in natural sciences. It is precisely this which is meant by the primacy of practical as opposed to theoretical reason.

At the same time, Kant, did not want to revert to the pluralism of value hierarchies determined by the existence of various integrations. He, too, thought in terms of humanity as totality. If natural sciences are capable of becoming the common language of the entire human kind, then only a morality that is similarly the common language of humanity could be placed above. That was how he arrived at the categorical imperative which abstracts from all concrete demands, value hierarchies, habits and inclinations just as the theoretical attitude of natural sciences does, and which, as a consequence, can be regarded to the same extent as valid for the entire humankind. The common language of morality postulating that a human being should not become a mere means for the other, attains to primacy over the language of natural sciences. To put it in a simple form, the natural sciences may consider nature to be a mere object, may manipulate it just as they want, as long as science does not thereby degrade human beings into a mere means. If it does, it collides with the other common language of humanity which has a constant primacy over the language of natural sciences.

Ever since it has become manifest how productive it is from the viewpoint of technical development to set nature as a mere object, how instrumental rationality founded on this basis could lead with ever-increasing speed to newer useful inventions, natural sciences cease to recognise their truth merely in the 'spell' of their own emancipation. Instead they set out to wage an aggressive war of conquest against all types of *epistēmē* which deverged from them, against all types that assume an independent form. It is precisely this aggressive expansion that is generally called positivism. It is not positivism if natural sciences (and mathematics) speak their own language, but it is if this language is postulated by them to be the sole and exclusive scientific one, and if they try to impose it on the understanding of the human social world and on social understandings, concepts of truth, and objectivity. It should be added to this that it was the philosophers rather than the natural scientists who played a leading part in that campaign.

Let me repeat that what characterized the new language of natural sciences was the setting of nature as a mere object and its consequent manipulation, a de-anthropomorphized concept of truth as the leading value, the abstraction from the concrete integrations, societies, values and ideas, from the 'special languages'.

In such an approach, the adaptation of the concepts of truth and objectivity of natural sciences (and therefore their attitude) to human society meant also a setting of society, historicity, human beings as mere objects to be manipulated. I would oppose two counter-arguments to this tendency: (a) this procedure is an impossible one – which should not, of course, exclude its being postulated as a norm, as a practical idea; (b) to postulate it as a norm is the most *negative* Utopia ever conceived by the human mind.

(a) Nature *can* be set as a mere object: the setting subject who does this setting is humanity which has no interests to defend. Since it has no interest conflicts with any other integration, to see nature as an object is not perforce ideological. If humanity is, however, rendered by our attitude into a mere object, it will by the same act be divided into two parts: the setting subjects (the scientists), and those who are being set (who are manipulated). In that case, since we have to deal with two groups opposed to one another, the setting subjects and those being set who have interests and interest conflicts, accordingly this operation has an ideological character. That which is, however, ideological cannot be — *ipso facto* — objective. Precisely in the sense of the interpretation of modern natural sciences, its concept of truth cannot be in correspondence with that of the sciences. Ergo: society (history) simply cannot be set as a mere object.

(b) The second counter-argument claims that to postulate society (history, humanity) as a mere object and hence fit for manipulation is *the most negative Utopia*, since it bereaves its object (humanity) of its very essence. It is precisely the human species essence of consciousness and freedom which it abstracts. Let me recall the Kantian formulation: this attitude means no more and no less than the setting of humanity as a mere means to be manipulated. Since I share the Kantian idea of the primacy of practical reason, I refuse this Utopia.

Let me recall, however, the essence of the Kantian theoretical proposal: he placed the universal language of morality above the universal language of natural sciences. This leaves a question unanswered: is an *epistēmē* oriented towards the knowledge of society (a philosophy, a type of social science) that abstracts from all concrete integrations, societies, values and ideas, from all the 'special languages' and speaks one common (symbolic) language, possible and desirable?

The possibility of a science of this type cannot logically be excluded, though taking into consideration the future within our reach it can be excluded practically. It is for the latter reason that I would rather examine the second problem: is such a science desirable?

Since they emerged, both social sciences and philosophy have constantly assumed the social scientist or the philosopher to have private-individual value preferences. In communal (relatively closed) societies these personal preferences menat the re-interpretation of values, while since the emergence of bourgeois society they meant the individual re-choice simultaneously of principle values and value hierarchies. An individual relation of that type to values is a precondition of the socially-oriented *epistēmē* being *evocative*, it is a precondition that it should speak to us, that it could influence our behaviour, that we could enter into communication with it. The loss of that function would bereave the socially oriented *epistēmē* of its essence.

Second: the value universe of every socially oriented *epistēmē* expresses some kind of affinity to the needs of a given society, a social stratum, a class. These needs orient the *epistēmē* not only in its choice of the subject matter, or its selection, but also in the understanding of the subject matter, in the elaboration of the ordering conceptual framework: they are inherent parts of the work or thought itself. To express the affinity to the needs mentioned above is precisely one of the functions of social theories which would be lost by accepting the sole and exclusive symbolic language.

A plausible counter-argument to this would be as follows: If we assume in the future a unified humanity, not stratified into social classes, we have to assume as well that this affinity, so crucial until now, will lose its significance. My reply to this is that even if we assume a non-stratified humanity in a future out of our reach, we should not assume one void of the pluralism of different ways of life, i.e. of the pluralism of the world of needs. The early Utopia, positive as it is in itself, turns into a negative one if it should be linked with a homogeneous way of life that cannot be any further chosen and re-chosen, since this would bereave humanity of the principle value of the positive Utopia of freedom. As a consequence, to refuse this Utopia is to postulate the plurality of different ways of life even in societies non-stratified into classes; hence we also postulate the plurality of the socially oriented *epistēmē*. We can no longer postulate a unified social science, but social sciences communicating with each other in debate each of which expresses affinity in its system of values to some particular way of life.

Let me now return to Husserl's theoretical proposition. The question is whether it is desirable to reconsider the attitude of modern natural sciences which calls for the re-creation of a unified human science, a unified *epistēmē*, a unified attitude to be the solution of the antinomies of present natural sciences. My question, then, is not whether modern natural sciences belong to history or not, since my reply to this does not deviate from that of Husserl — that is, I answer it in the affirmative; nor to whether modern natural sciences represent the only scientific form of knowledge pertaining to nature, since my reply to this — in accordance with the Husserlian position — is negative.

I simply ask whether we should opt — from the viewpoint of the human future — for giving up that historically developed form of knowledge?

What follows explains why my answer to this theoretical proposal is negative.

As demonstrated above, it is neither possible nor desirable to grasp society as a mere object. It has also been demonstrated that the formation of a unified symbolic language in social sciences would be equivalent to the revocation of freedom. Should we prefer, then, a unified social science, and hence thereby the annulment of the special attitude of natural sciences, it would lead to the following consequences.

On the one hand, we could not even render nature into a mere object (and manipulate it accordingly). However, as demonstrated above, it is precisely this manipulation that modern technique is based upon; its conquests are due to it. Various standpoints are possible and legitimate with respect to whether the pace and dynamics of technical development are adequate to the needs of humanity, and also with respect to what areas of social life are affected by the technical inventions. But I strongly doubt that there are thinking persons with the intention of destroying the attitude that can lead to such inventions in general. I think I express the explicit needs of the majority of humanity in regarding such a tendency as value destroying.

On the other hand, since the creation of a unified symbolic language is not desirable from the viewpoint of social sciences, it is rather questionable to give up the only possible symbolic language that can be spoken by the whole of humanity without thereby endangering the value of freedom. Apart from technical evolution, in my opinion this is undesirable either historically or morally: such a renunciation would vitiate a value which is historically already developed.

Melting the individual languages of natural sciences into a unified science would consequently be in both senses a negative Utopia irrespective of whether we try to impose the concepts of truth and objectivity of natural sciences on social *epistēmē* or if we try to impose the attitude of the social *epistēmē* on natural sciences.

Was it, then, sensible to pose the question at all? Does it not end up by saying: let everything be as it actually is? This is not to be inferred of necessity from the above premises. We cannot leave out of consideration (which can at the same time be regarded as an alarm signal) the need formulated by Husserl and his followers. "Something is amiss" with modern natural sciences.

We need not however derive from Galileo, from the creation of the unified symbolic language, from the new attitude, that which is "amiss". Our question is rather a reverse one: does modern natural science live up to the commitment it 'promised' to fulfil with Galileo, at the beginnings of the new attitude?

The new attitude of modern natural sciences was a gesture of emancipation: "to read in the open book of nature" also meant that the scientists abstract both from the *sensus communis* of concrete integrations and from their worldly powers in their researches. A unified language meant at the same time a commitment to humanity and to nothing but humanity. It meant that science, following its own logic, would not let itself be bothered by the logic of mundane powers. However, the natural science of our age does not fulfil that promise. As soon as the technical, industrial 'profit' of the new science became plausible, natural sciences became to an ever-increasing extent servants to worldly powers. Science is able to follow its own logic to an ever-decreasing extent; it has to serve power structures. It is precisely the remarkable scientists who know and feel the most that the orientation of their activity is opposed to the norm of their own science.

The language of modern natural science has developed as the common symbolical language of all humanity. This language cannot, however, be spoken by even the overwhelming majority of those being 'employed' in natural sciences; they are only acquainted with some of its linguistic games (Sprachspiele). The army of the skilled workers of physics and mathematics does not read any more in the "open book of nature" but obeys instructions without knowing to what totality of researches the instructions pertain nor what purposes they serve.

The attitude of modern natural sciences presupposed and developed 'publicity': all scientific results are public affairs, they pertain to everybody, every one can use them as a source of further knowledge.

However, modern natural sciences also renounced this promise. 'Secret' researches constantly grow in number. There are 'military secrets' and 'industrial secrets'; moreover, there are 'private secrets' too. In the struggle for life between the scientists, those who are going to win are those who can keep their secret ever longer to themselves and who grasp the moment correctly when the concealed result should be made public for the sake of a career. As far as attitude is concerned, we have returned to the age of alchemy.

So something has gone wrong with natural sciences. But that which is wrong does not derive from the original attitude but from the circumstance that it is only instrumental rationalism that has been left of the original attitude. The social-historical value orientation once represented by this theoretical attitude, which belongs to its norm, without which it is not what it ought to be, is step by step becoming wasted.

This losing ground of the original value orientation of natural sciences and its gradual disappearance is, however, a historical problem again and it can only be reversed in a hypothetical social-historical turn. It is precisely the social-historical preconditions of this 'reversal', toward which philosophical-sociological thinking ought to be oriented, not toward the annulment of the theoretical and value orientation of natural sciences.

My goal was to raise the problem, not to formulate a theoretical proposal for a solution, so I refer only to two problems in this respect. First of all: the natural scientist is bound in a two-fold sense in choosing his or her field of investigation and subject matter: by the logic of the science and by the manifest social need for the research. With respect to the latter problem the question arises: who or what represents the social need? In the first two centuries of the history of natural sciences the latter meant no crucial problem at all: the need had been 'touched upon' and grasped by the natural scientists themselves. Today, however, when scientific researches are realized within a powerful organizational framework and they need enormous investments, the personal 'grasping' of the need is no longer in itself sufficient: a social assignment must be added to it. Until there exist social institutions in which the social assignment of a concrete community or even of all humanity can become manifest, there is no general possibility for natural sciences to recur to their original value norm, they have to serve the powers which finance their researches. Second: we have to constantly bear in mind the Kantian postulate, the primacy of practical reason. It is a duty of the natural scientists to refuse all social needs directed towards researches the result of which is the use of one human being as a mere means for the other. They must abstain from researches of that type even if it is required by the logic of the science itself. This latter postulate is independent of the possibilities of any transformation of social structures in the future since in this respect 'here and now' prevails as a moral imperative. There are not two separate moral laws, one for the scholars and one for the laity.

Reference

1. Jean-Pierre Vernant, Mythe et pensée chez les Grecs: études de psychologie historique, Paris, F. Maspero, p. 32; and *passim*.

GUARDIANS AT THE FRONTIERS OF SCIENCE

INGO GRABNER, and WOLFGANG REITER (−1)
Universität Wien

Dedicated to Kunibert Knoppel (0)

1. Introduction

The hegemony (1) of science, which has, for several centuries, served as a frame and signpost to the development of western society, is now increasingly challenged. The questioning of political decisions in the name of scientific necessity without democratic process; the growing discomfort with the conditions of living in an overtechnicized society; the resurgence of 'irrational' practices and systems of belief; all these symptoms indicate that we may be witnessing a turning of the tide.

The following analysis is centered around the ideological significance of what are often and easily called the 'frontiers of science'. We shall extend this metaphor in order to ask further questions: if there are frontiers, what is beyond? Who are the guardians of these frontiers, and what are their aims? How is life at the front?

Our point of departure is that the white patches on the explorers' maps were almost never voids, but territories occupied by other cultures. In the same way, the frontiers of science are not the borderlines between knowledge and ignorance; rather, problems newly taken up by science invariably lead to questions to which other forms of knowledge or belief have already provided answers. These alternative approaches may or may not claim the rank of science; conversely, they may confront science with problems and solutions which it is not, or not yet, able or willing to adopt. In any case, the question arises how the demarcation is made between science and other systems of investigation. We aim to examine the criteria for this demarcation in the vast field of the so-called 'pseudosciences' — a deprecatory term coined by science itself. However, we shall not try to restore some, or all, of the pseudosciences to a respectable scientific place by proving that their methods and goals conform to the criteria set up by science; this would be a dull and fruitless task. Instead, we shall argue that questions typical for frontier science make the validity of the official criteria of demarcation very doubtful.

Admittedly, this step catapults us headlong into another confusion: in order to understand, let alone criticize frontier science, must one not have a profound knowledge of it? Indeed, the esoteric character and the formal difficulty of the foremost exponents of frontier science, particle physics and astrophysics, provide them with an almost impregnable rampart against criticism (2). This diplomatic immunity is reinforced by the philosophy of science: science, if at all, may only criticized from 'within'. The non-expert — and this applies to the expert in a neighbouring field of science almost to the same extent as to the layman — can overcome this interdiction only with a good deal of insolence. But let us not forget: the scientist generally does not hesitate to speak about questions which are completely foreign to his fields, e.g., politics; and he does not refrain at all from criticizing pseudoscience without being an expert at it. This attitude will serve us as a guideline.

The confrontation between science and pseudoscience will lead us to the core of our argument: demarcation of science against other systems of knowledge is a reality, and sometimes a violent one. If the criteria of science themselves are not applicable, why, then, must the frontiers be guarded? The key to this problem lies in the fact that science assumes the rank of a myth. We shall advance some conjectures about possible answers, which are as many attempts to locate 'anti-science' on the ideological level, relatively to the value system inherent in science.

Two more warnings for our reader: first, it must be remembered that we are dealing with frontier science only, not with scientific or technical undertakings taking place behind the front. We are well aware that the latter represent the main share in the economic potential of scientific research and development. This means that our analysis cannot yield any result whatsoever about the status of science as a productive power. The real importance and impact of frontier science lies precisely at the ideological level: inasmuch as it figures as high priesthood among a much more numerous clergy.

Secondly, we certainly do not claim to be able to present a thorough analysis of all these problems. Our approach is vague and fragmentary; an attempt all the more difficult as we are challenging long-established beliefs in the superiority of science. For anyone who has been socialized within the rules and values of science, this challenge is something of a parricide. We cannot spare the readers the task of drawing their own conclusions, from their own experience. We feel very strongly that a critique of science aspiring to a certain autonomy needs above all an overall picture of its scope, however self-contradictory and ambiguous this picture may still be. A critique of science seeking to be more than yet another subdiscipline of sociology or philosophy has no choice but to live and express all the contradictions which shake science itself.

2. Frontiers of Science

The concept of 'frontiers of science' is probably as old as science itself. Indeed, science has always been viewed by its proponents as an undertaking aimed at extending its domain to ever new fields of knowledge. This is probably why the term, although never defined with respect to any methodological or sociological theory, is nevertheless as easily understood — intuitively — as it is widely used in the practice of scientific research. As a more or less clearly definable discipline however, frontier science is a product of the explosive growth of science, and of physics in particular, which originated with World War II. Ever since, the investigation of those aspects of nature where new phenomena were to be discovered, new laws to be established, has evolved into a multimillion-dollar enterprise employing thousands of scientists. Frontier science, as an endeavour both theoretical and experimental, is identified with a number of prestigious scientific laboratories, the vanguard being the huge particle accelerators.

At the same time, particle physics — also termed high-energy physics — and astrophysics/cosmology are generally recognized as the most valuable and noble of all scientific disciplines. 'Science, the Endless Frontier' was the title of the 1947 report by which Vannevar Bush, one of the main organizers of the Manhattan Project during the war, urged the US government to supply academic basic research, especially in physics, with generous funds without interfering in the autonomous course of the development of science.

It was the commitment to basic research, a commitment of religious intensity, that was driving the leadership, and a large portion of the scientific community . . . (3)

This commitment has, if anything, only increased up to the present day. A collection of quotations taken from a small brochure entitled 'Nature of Matter. Purposes of High-Energy Physics' (4) may serve to illustrate some aspects of the world-view of frontier scientists:

The frontier of intensive research has always attracted a certain group of very clever scientists. To work in an uncharted field, to discover new laws of nature and completely new types of phenomena, is a great lure for a scientist. One is placed at the spearhead of a great and successful tradition ranging from Galilei, Newton and Maxwell to Einstein, Bohr, Dirac and Heisenberg. (V. Weisskopf)

Each human society excels at a small number of the many activities that people carry out . . . It is therefore an expression of the highest spirit of our culture to carry on with the task we have begun, the exploration of nature to all its limits. (G. Feinberg)

A great society is ultimately known for the monuments it leaves for later generations. We cannot foretell what detailed results may come from a very high energy machine . . . We can foretell, however, that such a machine . . . will without question be a source of inspiration for new science and a monument to our days. (A. Pais)

Thus . . . research in experimental high energy physics can play a modest but not entirely negligible role in maintaining economic stability. (H. Primakoff)

But all pure research is just a big marvellous gamble, in fact the only gamble so far invented that really makes sense. (G.C.Wick)

The world view of the physicist sets the style of a technology and the the culture of the society, and gives the direction to future progress. (J.Schwinger)

This little chat among some of the leading physicists of our time reveals the direction of their commitment: while not unrelated to the world of technology and economics, the main impact of frontier science is on the cultural level; here, it is truly basic.

Since forces drive all processes in nature, physics, as the science studying these forces, can be considered as the most basic. Without the discoveries that have been made in physics, no observations in any other field of science could be truly understood. If research in physics were slowed down or stopped altogether, all other sciences would have to suffer eventually, with immeasurable consequences for civilization now and later. (5)

Physics is a link in a chain of human efforts. If a link is broken, a chain is worthless. The frontier field of high-energy or particle physics, therefore, must be continued until man's search for understanding his environment has been successful. (6)

Nor is there any inherent limit to the triumphal advance of science:

It is reasonable to predict that no observable phenomenon need be considered alien to, or beyond the reach of, science . . . (7)

It is interesting to note the vocabulary used on such occasions: at times it seems to be taken from Don Fernando Cortez' notebook. "Invading new fields is a regular habit of ours", says physicist P.M.Morse (8). "All of nature and all of culture, including science itself, can be made to fall under the domain of science", according to the philosopher M. Bunge (9). Indeed: "The successes of the scientific approach . . . account for the expansive power of science, which now occupies territories previously occupied by the humanities — e.g., anthropology and psychology — and is continually exploring new territories" (10). The conquest is well under way: "I firmly believe that the same quantitative techniques that lead to the discovery of new facts of nature can be employed with appropriate modification to optimize the societal benefits of application or non-application of these facts . . . We physicists know about physical constraints; we can help society accept social restraints . . . I see no reason why we should not recognize Social Physics as a part of Applied Physics" (11). And the philosopher gives his final blessing: "Should anyone hold that the scientific approach does have inherent limitations we would ask him to sustain his claim — by conducting a scientific investigation of the problem" (12).

These slogans of a scientific colonialism show well that the purpose of frontier science is not just to "find unexpected and mysterious objects" and to "penetrate into the deeper and darker realms of the universe", as V.F. Weisskopf puts it (13). Actually, frontier scientists feel a heavy responsibility on their shoulders:

Without particle physics, physics would have no true frontier, and in a larger sense, science would have no frontier. Since no truly new principles or laws of nature could be discovered any longer, all science would have soon to become 'applied science' and ultimately new developments in technology and education in science might cease, and thereby education in scientific method, in objective discipline, and in abstract thinking, all of which are so badly needed to help in bringing about the required profound changes in human attitudes. (14)

Thus frontier scientists see themselves not only as conquerors, but also as defenders and promoters of a system of values which would collapse if science was not given adequate support — this, at least, is their apprehension.

Hence, they assume the double role of the warrior and the missionary. The myth of science has always had its preachers, and it needs them now more than ever. For sure, the preachers of science have always maintained an ambiguous position towards their colleagues and competitors in the religions — fighting them at times, searching for a compromise at others; God is an awesome enemy. Luckily, there is a more familiar enemy at hand: the Devil is never far off.

The World, the Flesh, and the Devil; an Enquiry into the Future of the Three Enemies of the Rational Soul, is the full title of Bernal's first book . . . Bernal saw the future as a struggle of the rational side of man's nature against three enemies . . . The third enemy he called the Devil, meaning the irrational forces in man's psychological nature that distort his perceptions and lead him astray with crazy hopes and fears, overriding the feeble voice of reason . . . (15)

And indeed vigilance is badly needed:

We are seeing among educated people a resurgence of superstition, extra-ordinary interest in astrology, palmistry and Velikovsky; there is a surge of rejection of rationality, going far beyond natural science and engineering. (16)

. . . It is especially disheartening to see the antirational and antiscience movement taking hold among our young people. Perhaps it is no overstatement to say that the problem with the highest priority of all is how to persuade our youth that preoccupations with witchcraft and astrology or the psychic experiences of the drugged brain are 'cop outs' in lieu of really doing something about the difficult problems that both generations agree exist. (17)

So, let us take a closer look at this terrible enemy. Where is the Devil, and how shall we recognize Him? (18)

3. Prospect from the Watchtower

3.1. A Look Beyond the Frontier: The Many Faces of the Devil

The Devil had a predilection for roving about in monasteries. Because there was particularly great harm to do there? Or because the pious monks had such a naive mind? or such an acute flair for Him? In any case, the monks' slang of the 13th and 14th centuries indulges in visions of the Devil . . . (The devils) appeared to them as horse, dog, cat, bear, monkey, toad, raven, vulture, or dragon, frequently even in human shape, as succubus to women, as succuba to men. . . (19)

Neither should we expect to be able to define *our* Devil once and for all. Let us call Him Pseudoscience, that at least gives Him a proper name, even if He will not enjoy the exorcising contempt implied by this term. A closer look will reveal the following possible incarnations:
— Theories opposed to accepted theories of science, e.g. freak cosmologies, Velikovsky, Lysenko, or pyramidology;
— Practices not backed by science, e.g. dowsing and various medical treatments, from acupuncture to homeopathy;
— Objects of inquiry not accepted by science, e.g. UFOs, the occult, parapsychology;
— Political movements, e.g. the ecological movement;
— Disciplines trying to develop their own methods not based on those of science, e.g. psychoanalysis or social anthropology;
— In some cases even scientific disciplines whose theories are still in an embryonic state and which are, therefore, mainly occupied with collecting data without being able to interpret them — e.g., biomagnetism.

All these various undertakings have been, and to a great extent still are, derogatorily labelled pseudosciences. Their baffling diversity raises the question: what the Devil do all these 'fads and fallacies' have in common? The only immediate answer is that pseudoscience is negatively defined as being exterior to science:

A body of beliefs and practices whose practitioners wish, naively or maliciously, pass for science although it is alien to the approach, the techniques, and the fund of knowledge of science. (20)

Pseudoscience, just like the Devil, is thus not a clearly definable and consistent phenomenon, but rather a code (21): one and the same principle shall be recognized through the most incongruous representations.

In view of the phenomenological diversity of the pseudosciences one would expect that formulating clear criteria of demarcation on another level would be an important task for those scientists who consider pseudoscience as a threat. As method is the basis for the official characterization of science, we should be able to count upon methodological rules of demarcation. But strangely, very few scientists have ever gone beyond an attitude of emotional rejection. Nevertheless, there are some attempts to formulate explicit criteria. We shall briefly examine two of them.

M. Bunge, in his treatise on the philosophy of science, devotes a short section to pseudoscience (22). He characterizes it as follows:

(1) Pseudoscience refuses to ground its doctrines and could not do so, having totally rejected our scientific heritage;

(2) Pseudoscience refuses to test its doctrines by experiment proper; moreover, it is largely untestable because it tends to interpret all data in such a way that its theses are confirmed no matter what;

(3) Pseudoscience lacks a self-correcting mechanism: it cannot learn from either fresh experimental information (which it swallows without digesting it), new scientific discoveries (which it despises), or criticism (which it rejects indignantly);

(4) Pseudoscience has, like magic and like technology, a primarily practical aim rather than a cognitive one, but, unlike magic, it presents itself as science, and, unlike technology, it does not enjoy the backing of science.

Our second example is not taken from a scientific work, but from a book intended for the broad public: *Fads and Fallacies in the Name of Science* by M. Gardner, annotator of *Alice in Wonderland* and for a long time editor of the 'Mathematical Games' column of the *Scientific American* (23). Gardner, who divides his cranks and charlatans into twenty-six different categories, does not bother to formulate a single methodological criterion; instead, he obviously assumes that once some scientific authority has condemned a particular activity as pseudoscientific, the matter is settled on this level. But as this is not the layperson's level, and as the public should learn to discern between a real expert and a false one, Gardner provides us with some further clues:

(1) The modern pseudoscientist stands entirely outside the closely integrated channels through which new ideas are introduced and evaluated. He works in isolation.

(2) Pseudoscientists have a tendency towards paranoia: (a) they consider themselves geniuses; (b) they regard their colleagues, without exception, as ignorant blockheads; (c) they believe themselves unjustly persecuted and discriminated against; (d) they have strong compulsions to focus their attacks on the greatest scientists and the best established theories; (e) they often have a tendency to write in a complex jargon.

Strikingly enough, neither set of criteria can be said to be a methodological one. Gardner's characterization of the pseudoscientist clearly remains on the socio-psychological level. Also, Bunge's criteria stray onto this level more than once. The pretended self-isolation of pseudoscience, its presumed attitude towards criticism, the backing of technology by science, all this cannot be explained by purely methodological arguments. Moreover, since both sets of criteria do not aim to define pseudoscience in an abstract way, but must be applicable to all its concrete manifestations, they run into various contradictions. Whereas Gardner explicitly mentions attacks against established theories, Bunge notes the absence of cognitive aims — and chooses psychoanalysis as one of his examples, whose cognitive importance cannot be denied — however deplorable one may find that (24). Another equivocal point is the self-isolation of pseudoscience; it is somewhat daring to reproach it with the desire to be recognized as a part of science, and to assert at the same time that it places itself deliberately outside and against it. In fact, attacks against science as a whole very rarely come from the pseudoscientists, who at the worst will get excited over some scientific authority whom, rightly or not, they consider to be their enemy. Most pseudoscientists would firmly deny that they want to make a total break with the heritage of science; they may be opposed to entire theories, but they would never betray the myth of science. Gardner's first criterion, therefore, is a mere tautology: pseudoscientists are isolated because the scientific community does not accept them. As far as points (a), (b), and (e) of Gardner's second criterion are concerned, we leave it to the reader to falsify their occurrence within the scientific community.

The methodological content of Bunge's criteria is reduced to two conjectures: pseudoscience refuses to test its doctrines; and it lacks a self-correcting mechanism. These are doubtlessly the most dangerous traps for any scientist working in isolation, no matter whether this isolation was chosen or enforced. Therefore, these criteria will, of course, be useful to show up one or the other crank; but this is not a problem of demarcation. It is very doubtful whether they can be applied to pseudoscience which, like science proper, is characterized by merciless competition and the personal ambition of its exponents. Bunge's criteria do, therefore, not bring us any closer to a concrete methodological demarcation of those disciplines which are actually called pseudosciences.

We conclude that the demarcation drawn by Bunge and Gardner mainly serves to corroborate a state of social affairs, and much less to protect truth against untruth and nonsense. It is the vindication post rem of a code of which we cannot say more, for the moment, that it seems to be of great importance in the practice of scientific research.

3.2. Goal and Method of Science: The Devil Inside

Before we ask ourselves what the actual significance of this code is, let us remain for a while in the luminous domains of methodology. According to Bunge, the doctrines of pseudoscience are largely untestable. We would thus expect science, and in particular its frontier disciplines, to be especially scrupulous about putting its hypotheses and theories to test. Evidently, science is unthinkable without the experimental method — no pseudoscientist would deny that. But are questions without relation to experimental situations actually banned from science? (25).

Undoubtedly, scientific hypotheses are not created by the rules which the methodologists reconstruct afterwards. Ad hoc assumptions and intuitive insights play an important part, though this state of affairs makes philosophers of science somewhat uneasy and they tend to relegate it into the twilight of psychologism.

An essential criterion for the practical usefulness of a hypothesis is its plausibility — in physics, for instance, its 'physical content'. These are vague concepts which the methodologists would rather avoid. In any case they relate to the conformity of a hypothesis with the experimental method:

> The plausible hypothesis is a reasonable conjecture that has not passed the test of experience but may, on the other hand, suggest the very observations or experiments that will test it: it lacks an empirical justification but is testable. (26)

That a conjecture is 'reasonable' means that it does not contradict basic principles experimentally well corroborated. Therefore, plausibility is denied to many questions and hypotheses of pseudoscience; e.g., extrasensory perception is said to be incompatible with the principle of causality and difficult to reconcile with the principle of conservation of energy. To this, the honest parapsychologist can only reply that his hypotheses are, if nothing else, indeed based on well corroborated experimental observations; after all, observations cannot simply be dismissed because they seem to contradict some basic principle of science. That leaves the scientist without argument; indeed, "if theories are the end-product of science, experiments are the driving force" (27).

And so nothing remains for the critics than to declare a petty war on pseudoscience blaming it for negligence or even conscious fraud in its experiments (28).

The scientists, who obviously like to pose as Knights of the Holy Grail of Experimental Method, are, on the other hand, not averse to relying on other driving forces besides experiment:

> There is an unwritten precept in modern physics, often facetiously referred to as Gell-Mann's totalitarian principle (29), which states that in physics 'anything which is not prohibited is compulsory'. Guided by this sort of argument we have made a number of remarkable discoveries, from neutrinos to radio galaxies.
> Several such searches are in progress now. Because theory does not exclude the possibility that a magnetic analog to the electric charge can exist, physicists persist in their quest for the magnetic monopole. A similar search is on for the 'quark', a fundamental particle having 1/3 of electronic charge, whose existence is suggested by recondite symmetries of elementary particles.
> In each of the above cases, the possible existence of the new particle was first suggested by the logical extension of the regularities or symmetries governing the known physical world. The special theory of relativity provides for just such an extension. We have called it 'meta-relativity'. (30)

In the above-mentioned case, the result of this procedure is the theory of particles that travel faster than light, so-called tachyons. While their existence cannot trivially be ruled out, yet there is no experimental observation whatsoever pointing to it — nor is it at all clear how tachyons should interact with subluminal matter. This lack of experimental evidence does not seem to prejudice the hypothesis, although it is very doubtful whether it is testable and therefore, methodologically speaking, plausible.

We must conclude that it is apparently prohibited to ask questions about phenomena which are not predicted by the formalism of science; but it seems to be allowed, even compulsory, to extend a formalism to domains where one cannot be sure that there will be any phenomena. Is physics, then, not about facts of nature, but about mathematical structures? "I think it is more important to have beauty in one's equations than to have them fit experiment ...", P.A.M. Dirac once said (31).

But the problem is not that simple. One could certainly speculate that pseudoscience would fare much better if it first constructed beautiful mathematical theories and later interpreted some results to prove, e.g., the existence of psychokinesis. But this is not the whole point. The beauty of formalism cannot save a theory if it is easily refuted by observation. This, however, is exactly what is not likely to happen very soon to tachyons, or to Mr. Higgs' bosons (32). We must conclude that in frontier science it is good tactics to make predictions which are *in principle*, albeit not very easily, testable by experiment, i.e., whose untestability cannot be logically proved.

A transcendent (infinitely fast) tachyon should be particularly difficult to detect, as it has no energy or momentum to spare . . . (33).

This unconditional confidence in the explanatory potential of mathematical formalism is probably not unrelated to the physicists' difficulties with the concept of physical reality. This has been a central point in the discussion about the epistemological basis of quantum theory, which is still far from settled (34). On the contrary, physical reality seems to evaporate at an increasing rate in the rarefied atmosphere of frontier science. The mysterious 'observer' of quantum effects is making a remarkable career in cosmology and has lately been promoted to the status of 'participator', without whom the universe is meaningless:

The quantum principle . . . demolishes the view we once had that the universe sits safely 'out there', that we can observe what goes on in it from behind a foot-thick slab of plate glass without ourselves being involved in what goes on. We have learned that to observe even so minuscule an object as an electron we have to shatter that slab of glass . . . We have to cross out that old word 'observer' and replace it with the new word 'participator'. In some strange sense the quantum principle tells us that we are dealing with a participatory universe. (35).

In particle physics, on the other hand, as a consequence of the great deception of the unsuccessful quark-hunt, and in order to preserve the beauty of the equations (and the brand new 'charm' quark), it became necessary to explain why it might after all be utterly impossible to find a free quark:

Quarks are a product of theoretical reasoning. They were invented at a time when there was no direct evidence for their existence. The charm hypothesis added an extra quark explaining the properties of another large family of particles when those particles had themselves never been seen. Color, a concept of even greater abstraction, postulates three varieties of quarks that may be distinct but completely undistinguishable. Now theories of quark confinement suggest that all quarks may be permanently inaccessible and invisible. The very successes of the quark model lead us back to the question of the reality of quarks. If a particle cannot be isolated or observed, how will we ever be able to know that it exists? (36)

Such statements are likely to make some of our philosophers of science somewhat nervous, even if they do not belong to the empiricist breed (37). By the same token, we must concede that frontier scientists are not bothered too much by Bunge's second criterion of demarcation. At least it appears that the non-observation of a fact can readily lead to the extension of a theory which now aims explicitly at explaining the fundamental impossibility to make the observation. Therefore, if some day the theories of quark confinement were to yield testable consequences, it would be in spite of, rather than thanks to, the theorists' efforts.

Let us retain, then, that physics is not all that sure that its objects — and most certainly quarks *are* thought of as objects, since they are to be the fundamental constituents of all matter, and not mere mathematical auxiliaries — belong to the material world. Then why the indignation about 'spirits' in psychoanalysis, and the like? (38)

But frontier scientists may well ask far wilder questions, and indeed, why shouldn't they? There are popular themes of speculation like the origin of the universe, shared by top cosmologists (39) and the Catholic Church (40); or high-level international meetings about the problem of how to communicate with extraterrestrial intelligent beings (41); one may talk about the colonization of space (42), or discuss the "logical possibility that our galaxy has been colonized by extraterrestrial beings, and that we are being ignored, avoided, or discreetly watched..." (43).

Really, these are problems where the creative mind is free to roam about in breathtaking perspectives:

I assume (i) that angels are still within the Galaxy, and (ii) their average 'progress' can best be estimated as equal to ours ± millennia. Then a reasonable proportion will heave reached the stage of interstellar (wisely 'unmanned') travel — but not of being able to discriminate Earth from the multitude (the UFO seers are incredibly geocentric in their conceits). Ergo: if there is a transmitter, it will be at the unique point in the Galaxy — as far as I can imagine, the barycenter is just that. This will be even more self-evident to that subset of angels who live a few hundred or thousand light years from the center... (44)

Heavens rejoice!

A concerted national, and perhaps international, effort to find radio signals from extraterrestrial civilizations will probably be started within the next few years . . . (45)

At this point we cannot help feeling some sympathy for the poor outlawed UFOlogist — though we think that looking for UFOs on a cold winter night is a rather dull pastime. But still: "After all, the U in UFO only means unidentified. . ." (46)

3.3. A Second Look Beyond: The Magic Mirror

The frontier scientists obviously make good use of Feyerabend's maxim (47); and undoubtedly scientific progress could, if anything, only be hampered by methodological rigidity. But now we begin to wonder why the pseudosciences have to suffer such an outright rejection from the overwhelming majority of orthodox scientists. How come that frontier scientists claim for themselves all the advantages of the 'anything goes' slogan and at the same time, with a self-confidence that amounts to precognition, reject pseudoscience as rubbish?

But the interrelation between science and pseudoscience is still more complex — and tighter — than it may have seemed up to now. On the level of interpretations, they correspond to each other like mirror images, and the mirror, in this case, very likely is a magic one. Pseudoscience desperately tries to get some recognition by referring its conjectures and observations to what it thinks to be 'advanced' science — in particular, quantum physics; and science itself prepares the ground for this attempted identification by the interpretation of its own results. We have already seen some of these interpretational labyrinths in connection with the problem of physical reality; once engaged in them, science and pseudoscience will not fail to meet at some corner.

The desire to be admitted into the great womb of science is omnipresent in most disciplines which must manage to get along beyond the frontiers. This is understandable on the institutional level since most available funds are controlled by scientific organizations (48).

But beyond this need for a respectable appearance, most of the pseudosciences are surprisingly reluctant to develop their own, autonomous theories and methods. Some, like parapsychology, are at great pains to improve their experimental and statistical techniques beyond any possibility of orthodox doubt — mostly in vain, as very few scientists are prepared to engage seriously in that discussion. While the biologist, or the quantum chemist, hardly ever will doubt the repeatability of experiments in particle physics, even knowing that, in fact, very often they are never repeated, he will certainly contradict the repeatability of a parapsychological experiment without taking a second look at it.

Luckily, pseudoscience is not easily discouraged. The second attempt to chum up with science has proved to be more successful. Just as Freud, in his early days, expressed the hope that psychoanalysis would ultimately the reduced to neurophysiology, many disciplines concerned with the studies of occult phenomena are almost convinced that what they observe is in reality nothing more than another trick played by physics.

> To understand such phenomena (out-of-body experiences) may involve moving... into a reconsideration of the nature of time and space themselves. Coincidentally, these have become crucial questions in twentieth-century physics as well... It may well be that the advance of occult physiology and the next great advance in physics will take place together. (49)

The average scientist will just smile in astonishment at these extravagant hopes. But is science really innocent of them?

> In the bizarre universe of quanta and indeterminism, 'psi' forces... seem to insert themselves very naturally, at least according to certain physicists who are partisans of the Copenhagen interpretation of the Dirac equations. We even think that these physicists could provide us, probably soon, with the general theory which we lack so much; and this lack of theory, by the way, is the most serious criticism which could be made of parapsychology. (50)

What the Devil! (thinks the dumbfounded physicist) — the Copenhagen gang again?! Some bad reading perhaps? But no — these are conclusions drawn from the discussions held at a Very Respectable Conference on 'Quantum Physics and Parapsychology' (51) . . . Have you noticed how a subtle psychokinetic pull has drawn us to the other side of the magic mirror? Indeed: some guardians, intoxicated by their own pseudophilosophical dreams, are sadly failing to fulfil their duties.

Actually, this is a question which deserves better than a tongue-in-cheek treatment. J. M. Lévy-Leblond speaks of a "proper ideological exploitation of modern physics" (52). The reckless extension of a handful of erroneous interpretations of quantum physics to many an aspect of individual and social life has served to legitimate almost anything from human free will to the superiority of western liberalism. Clearly, the pseudosciences claim the fertile soil they find here; and they cannot even be blamed for it. For if quantum theory, according to P. Jordan, proves the existence of God (53), why should it not be qualified for proving psychic action? If physicists seriously contemplate such curiosa as transcendent tachyons, who can prevent parapsychologists from harnessing a couple of them onto a nice explanation of telepathy? Thus scientists themselves help pseudoscience jump the bandwagon of science.

Again, this is not the whole story. If it were only for some easily corrected false interpretations, no great harm would have been done. But how should the non-expert be able to correct anything presented by the expert as a true corollary of his scientific knowledge? The more frontier science withdraws from domains of physical reality accessible to everyday experience, the more its results must necessarily be accepted by the layman as a modern Revelation: Annunciation or Apocalypse (54). In this way, the secular power of the US President's science advisor acquires ever more features of the spiritual power of the High Priest.

This state of affairs is being acknowledged by some scientists, who begin to discern that equation of scientific progress with enlightenment is not self-evident:

Many must have been relieved to hear that only Einstein and a few geniuses are able to understand the world. They had tried to understand science the best they could. But now it became clear to them that science was something in which one was to believe, and not something which one should try to understand. In a paradoxical way, Einstein has perhaps not been cheered by the public because he was a great thinker but because he relieved the others from the obligation to think. (55)

Science has helped to shape the world in a way which makes it increasingly difficult for almost anybody to understand what is going on around them. Henceforth, a critical attitude towards science, which requires a certain acquaintance with it, becomes almost impossible. If science does not succeed in breaking out of its sanctuary, the only way for non-experts to deal with it will be to believe that anything done in its name is inherently right — or wrong. Indeed M. Gardner offers a fine example for the perversion of this absolute thinking, when he speaks of the "frantic efforts of the military to persuade a dazed public that nothing sinister or extraordinary was taking place above their heads. . ." (56). This was written sometime in the early fifties, and he meant flying saucers, of course — nuclear weapons, being the legitimate offspring of scientific research, can for that very reason be neither sinister nor extraordinary.

At the same time, science has monopolized the concept of rationality. Everything that is said or done in its name is ipso facto rational. There is no supreme authority entitled to judge it. Thus, science's claim for all-inclusiveness dissolves rationality at its frontiers, without, however, endangering its cognitive hegemony.

If it is not rationality which is to be defended at the frontiers of science, what, then, is it? To come nearer to an answer, we propose now a brief random walk across the socio-economic reality of frontier science.

4. The Guardians at Work

4.1. The Breeding of Dinosaurs

Perhaps the most conspicuous feature of frontier science is its extravagance: the budgetary volume and organizational complexity of research, especially in particle physics, dwarf all past efforts of gaining basic knowledge. Only the wealthiest nations can afford to approach the frontiers. In the USA, the annual federal support for high-energy physics lay, in the past ten years, somewhere between 0.15 and 0.3 gigadollars — most of this little pocket-money being divided between no more than five accelerator laboratories; older installations must inevitably be shut down after ten or fifteen years of operation. Once again, the conquest of the unknown has begotten a merciless competition in which only the most powerful nations can participate. The frontiers of science, however, are not necessarily congruent with national frontiers. Although factors of national prestige as well as of economic and political domination play a considerable role in frontier science, it is not a priori determined by national or imperialistic ideologies. The next great step in particle physics may perhaps necessitate a worldwide collaboration; the 'very big accelerator' that some physicists are demanding may be beyond the economic possibilities of any single superpower. As yet, the myth of science transcends the frontiers between political ideologies.

The temples of the myth are the big accelerators. Were it not for their gigantic dimensions, physicists would not see in them the "monuments to our days" or compare them with the Acropolis or with Gothic Cathedrals (57). Frontier science is the biggest cultural show of our times.

There is however, one major difference to other shows: the myth of science has no accessible place of worship; particle accelerators or radio telescopes are not erected in prominent places in the centres of metropoles. Physicists could never get support for their ambitious plans just by stressing their importance as a cultural avantgarde. And so, their second argument has always been that frontier science is fostering technological progress.

A point in case is the immediate application of particle accelerators to ends other than those of basic research. Most prominently figure biomedical research and application for therapeutical purposes; the magic word 'cancer therapy' has helped to silence many a doubt in the utility of investments in particle physics. Of course, the potential military use of accelerators is also emphasized from time to time; occasionally, some high-ranking general might be persuaded of that possibility, and, by applying his own implacable logic, convinced that the Russians, years ahead of the USA, have developed a new lethal weapon (58). In the same way, one may speculate on the possible use of neutrino beams for telecommunication (59), and so forth.

Nevertheless, this discrepancy between the gain to be expected from these applications and the necessary investments remains so considerable that they can be regarded mainly as an a posteriori justification of the construction of particle accelerators. A second argument must then be produced: frontier science needs and promotes frontier technology, which is a worthwhile undertaking *per se*. In addition, technological developments triggered by frontier research can again induce new techniques which lastly will benefit industry, the military, or even Mr. Nader's consumers.

This point contains some truth. Frontier science has given birth to an important secondary industry, thriving on the generously endowed federal commissions associated with the construction and operation of accelerators and other installations. Possibly, this expenditure contributes somewhat to economic stability by surplus absorption; this is, however, a minor effect, since these industries are not nearly as powerful as, say, the big aerospace corporations.

Once again, it seems that the impact of the argument rests mainly on the ideology which confers to frontier science a basic role for all aspects of society. The fundamental character of particle- or astrophysics consecrates the technology to which it is amalgamated into an order of basic and universal importance. Yet when this assertion is particularized, frontier technology emerges as a mixture of megalomaniac techniques and ridiculous 'spinoff' gadgets (60).

But this does not matter any more. The myth of science and the technological megalomania of modern industrial society support each other in mutual justification. The sermons held in the temples of frontier science tell us not only about truth and knowledge, but also which earthly goods are pleasing to God. There is a direct correspondence between Big Science and Big Technology, between the alleged need to build ever larger accelerators and the assertion that only huge, centralized industries can solve the production problems of contemporary societies (61).

4.2. Mandarins and Monks

The temples of frontier science are far removed from everyday life; they are not, however, uninhabited. Experimental research in particle physics and in other fields employs huge numbers of scientists and technicians, and is organized according to industrial standards. Achieving and maintaining a reasonable output requires tight organizational schedules and an efficient management.

The production process itself, on the other hand, is not industrialized. Particle physics is not tailorized, results are not produced on the conveyor belt. Quite on the contrary, it works somewhat on the principles of the ant-hill specialized in skilled craft. As a rule, one experiment involves great numbers of scientists (short publications by 20 authors or more are quite common), not to mention technical and administrative staff, and lasts several years.

Within these research groups, division of labour is thorough, founded both on hierarchy and competition. The traditional roles of experimentalists and theorists are complemented by those of the project managers; the division affects not only power, authority and competence, but also knowledge and information. Theorists initiate and outline a project, managers organize it, and the crowd of experimentalists realizes it, each according to their competence in which they try to make themselves as indispensable as possible.

The experimentalists occupy the lowest rank in the hierarchy. As novices or as ageing monks, they are part of a machinery whose purpose and operation they cannot influence, and of which they understand rarely more than their own part in the work. As of old, their lives follow the monastic principles of poverty, chastity, and obedience. One distinction has disappeared: sexuality has no place, women have to function as men. Life is simple, the rules clear, communication transparent and compulsory: technical constraints determine the course of machine-oriented days and nights.

On the higher levels of the monastic hierarchy, life is more complex. The proximity of the profane world can be felt, particularly the managers cannot avoid contact with it. The theorists carry the heaviest burden. They are responsible for the spiritual health of the enterprise, for the well-being of the doctrine and its proper application They are the exegetes — a function which the experimentalists hardly ever assume, even when promoted to managerial honours. Mission is unequivocally the competence of theorists.

There are thus different levels of truth and knowledge within this hierarchy, distinguished by varying degrees of depth and range. The highest ranks belong to the Mandarins of frontier science, very prestigious scientists, mostly theoretical physicists, heavily decorated, holders of tenure at the holiest temples, disciples of other Mandarins. By virtue of their position, they are entitled to the greatest possible knowledge, the deepest wisdom.

I felt that if I could understand theoretical physics I could understand anything. (62)

The Mandarin's voice, however, does not cry out in the desert. In most cases, he seeks the proximity of the temporal power, is a manager himself, maybe even a President's science advisor. He is not only a prophet, but also a sales representative of the miracles of frontier technology.

The power of the human mind to think abstractly, and yet at the same time to say something of relevance to concrete reality, is nowhere so clear as it is in theoretical physics. (63)

The mandarins of frontier science are Janus-faced. With the ecstatic face of the truth-seeker on one side and the grim face of the expert on the other, they achieve a flawless synthesis between the most abstract knowledge and very concrete political influence. It is this holy alliance which we now propose to analyze briefly and we consider the Jason Division as an example.

4.3. The Scientist Bless the Arms (64)

The history of the Jason Division is as well documented as can be expected from a largely classfied matter; it need not be reiterated here. Suffice it to recall that Jason is a group of distinguished scientists, varying in composition, carrying out research and advisory tasks for the U.S. Department of Defense under the auspices of the Institute of Defense Analysis. In the sixties, Jason organized a number of study groups dedicated to the application of new technologies in the Indochina war. Protests by students and scientists, and the publication of the Pentagon Papers made these activities known to a wider public.

Jason is a very exclusive club: only Most Distinguished Mandarins are admitted, preferably physicists — but by now the absence of such a restriction would surprise us. Among those whom we have already encountered on our thorny way through the labyrinths of frontier science, we shall meet again as Jason fellows: Gell-Mann, Dyson, Wheeler, Weinberg, and Lederberg. True, the last one is a geneticist — but, after all, a Nobel laureate.

This highly honourable brains trust was (and maybe still is) very well paid for advising the Department of Defense, among others, in the following matters: electronic warfare, new weapons, ABM, SALT, counterinsurgency, and other (occasionally also non-defence) matters. You may well ask what the most abstract thinkers far and wide could contribute to the solution of these problems?

The Jason members could not be said to share a political conviction. The Pentagon did not hire them for their martial spirit. Some of them were indeed 'Hawks' and did not deny it. Others, however, were more or less outspoken in their opposition to U.S. involvement in Vietnam or had a reputation for their oppositional stance in other matters. Furthermore, new Jason recruits were taken from within the scientific community on the Mandarin level; it was a matter between friends and colleagues and not of being a general's protégé. Most war physicists see themselves as peaceloving citizens — God forbid that they would get their hands dirty. They are loyal servants of their government and when asked for their opinion, they are no doubt convinced that they have advice to offer which is not only relevant, but sound and humane.

The question is if I am to feel ashamed or proud of what I have done. I am glad to state publicly that I am proud of it. If my work had no effect on government policy, I can have done no great harm. If my work had some effect, I can be proud to have helped to avert a human tragedy far greater than the one we have witnessed. (F. J. Dyson)

Under favorable circumstances one's leverage in strengthening one point of view can be enormously large when coupled with honest technological work... Young people often ask scornfully what positive good the 'defense intellectuals' have achieved — the fact that the human race is still functioning today may be such an achievement! (E. E. Salpeter)

We all believed at that time that the things we were working on would help to bring the war to a close. In this we were probably very naive... There weren't very many wise men around in 1966. A large number of our colleagues who sat around sucking their thumbs for years as well as a majority of the American people are equally worth of condemnation. (M. L. Goldberger)

It is not consistent to condemn those scientists who aid our national defense without condemning the majority of the US population... Is it right for a scientist to weaken the national defense by refusing to contribute his part to it, so long as this is a democracy?.. Such consulting introduces the thinking of university people into government planning. (A. A. Broyles)

At the risk of appearing sanctimonious, I must say that the basic issue seems to me the one raised in Luke, Chapter 6, Verses 30—31. "But their scribes and pharisees murmured against his disciples, saying: 'Why do ye eat with publicans and sinners?' And Jesus answering said unto them: 'They that are whole need not a physician, but they that are sick.' " As a member of Jason, I sit down with all kinds of people who are caught up in one part or another of the United States government and the armed forces. You can call them publicans and sinners if you like. They are people like us, facing difficult problems and badly needing contact with the outside world to give them a clearer perspective. As a Jason member I am given the opportunity to talk with these people and to bring them into touch with reality as I see it. I cannot know whether my attempts to influence their attitudes are successful. But I know that I would be betraying my responsibility to humanity if I did not try.

It seems to me that the position adopted by SESPA is close to the position of the Pharisees. People inside the government are sinners and the rest of us should preserve our purity by denouncing them. Jesus did not see it that way and neither do I. Of course it is ridiculous to make a comparison between the feeble efforts of Jason members to talk sense to the generals and the efforts of Jesus to change peoples' lives. But it is important, quite apart from Jason, to establish the principle that one may eat and drink with sinners without being used by them. Was Jesus used by the company he kept? (F. J. Dyson)

By and large, we see that our Mandarins are actually convinced of their missionary vocation, and of the support of a majority of the population. Consequently, they do not hesitate to throw themselves into the heart of the battle. Even E. E. Salpeter, who in 1968 officially resigned from Jason because he despaired of being able to achieve a slowdown, let alone an end to the Vietnam war, is still convinced that 'honest technological work' can have an 'enormously large effect', in this case nothing less than the salvation of mankind. If these statements may seem absurd to anyone not familiar with the ways of the scientific community, they are, however, consistent with the myth of science. The myth, being absolute, does not recognize the existence of problems that scientists could not solve better than anybody else.

We have two more questions to ask. First: what reality are the physicists talking about, what outside world should the wicked politicians come to know? The world of cosy campuses, of leisurely lectures, of heated debates among brilliant intellectuals about some scholarly question? The reality of scientific meetings held in the ever identical glass-and-concrete congress halls, of the de luxe hotels all over the world? Or the reality of Jason group meetings, held in a friendly and uncompetitive atmosphere in such pleasant places as La Jolla or Woods Hole, and for which participants were paid 100 to 200 dollars a day? What has the life of a Professor of Physics at the Institute of Advanced Studies in Princeton, or that of the Higgins Professor of Physics at Harvard University, in common with the life of those who are really affected by decisions about a war somewhere in the world, or counter-insurgency measures taken by the U.S. — again somewhere in the world? (65).

Secondly: were the Jason members used by the company they kept? Probably not in the sense Dyson meant, namely that they were forced to renounce some of their principles. But perhaps that was not even what the military expected from them. Perhaps they were content to get the Mandarins' blessing for their decisions. What can be wrong with a war whose methods were improved by the inspired ideas of the most prestigious scientists, the defenders of humanity on every front?

Scientists and technologists had acquired the reputation of being magicians who had access to a special source of information and wisdom out of the reach of the rest of mankind", writes H. York, himself an ex-Jason member, a long-time activist in science policy and today one of the most outspoken critics of the course this policy is taking. "Many of the scientists and technologists themselves believed that only they understood the problem. As a consequence, many of them believed it was their patriotic duty to save the rest of us whether we wanted them or not... Today there are even more people who tell us that because major weapons systems are so complicated only weapons experts can decide if they are needed, only those in on all the secrets and up to the most arcane elements of operations analysis analysis can tell us whether arms control and disarmament is good or bad, and only nuclear experts are fit to decide whether, when, and where nuclear plants should be built . . ." (66)

Thus political advice by scientists is no longer a question of who is used by whom. Just as scientists find it quite natural that their advice in political matters is sought, politicians find it quite natural that these matters are increasingly considered as purely technological ones. The ivory tower of pure science has nowadays been equipped with laser-guided weapons and surrounded with an electronic fence. The general and the High Priest share a working room: each one needs the other's power and at the same time fears it. Thus they provide each other with justifications of their respective hegemonies, cognitive the one, political the other.

5. The Myth of Science

5.1. The Criteria of Demarcation: Some Conjectures

But perhaps we have now ventured a little too far into the dark realms of politics, and the distinguished sociologists who must read this essay as part of their professional duties will find that inappropriate. Instead of an apology, we will insert a very dry and abstract, but altogether short, section that will certainly put our readers in a conciliatory mood.

What we have tried to demonstrate is that the methodological basis for the demarcation between science and pseudoscience is scant, to say the least. Nor does pseudoscience, from its own standpoint, wish for this demarcation. This leads us to the conclusion that the demarcation can only be understood with reference to the patterns of thought and action underlying the practice of scientific research, and the preceding three sections were aimed at exemplifying some of these patterns.

Extended to an ever growing domain of human activities within the frame of a world shaped and organized by science and technology, these patterns establish science in the rank of a myth. In the practice of scientific research, obedience to the myth is guaranteed by implicit precepts, whose violation is not just the failure to conform to the rules of a game; the outcome is not the irrational, but the ridiculous, the obscene, the unthinkable: anti-science.

Admittedly, the concretization of these contentions is largely a matter of future elaboration and discussion. Philosophers, of science and other subjects, have constantly and successfully avoided dealing with them. As a philosopher, it seems to us (but of course we are laypeople), you must either put a candid trust into the omnipotence of science, or else spell it as a four-letter word, say syns, and relegate it into one of the many dark corners of everyday life where philosophy does not set its celestial foot.

As for ourselves, we cannot do more than offer some conjectures. Principally, it does not bother us whether science describes reality, or constitutes it. Whatever the case, the fact remains that some, perhaps many, possible aspects of reality simply escape.

The myth of science rests upon three basic precepts: (1) science is all-inclusive and expansive; (2) scientific knowledge is a hierarchic unity; and (3) scientific knowledge is technically grounded.

The first of these precepts implies the assertion: no problem is exterior to science. Hence, there are only problems which can be formulated in terms of scientific and technical knowledge. Problems which cannot be so expressed, do not exist.

Furthermore, being all-inclusive, science and technology are neutral, and the same applies to those who possess scientific and/or technical knowledge and who are known as experts.

If science defines what can be known, it does not actually know everything. Scientific research is an expansive undertaking aimed at conquering new fields of knowledge. The frontiers of science cannot be stationary — on the contrary, newly acquired knowledge is often regarded as the most valuable and on the other hand, knowledge is subject to a process of ageing. Old knowledge may be forgotten and disappear, or it may be incorporated into common knowledge and no longer qualify as science. It follows that science cannot exist without expansion.

According to the second precept, scientific knowledge is hierarchical. To a hierarchy of natural phenomena corresponds a hierarchy of scientific disciplines, which, at the same time, is a hierarchy of values. Physics is the most basic, important, beautiful, and worthwhile of all sciences. Of course this preponderance of physics has benefited by the strong historical assistance of the explosive success of the Manhattan Project. But there are two further solid pillars it relies upon: the — at least theoretical — possibility of reducing all scientific knowledge to the basic laws of physics; and the unity of the scientific method.

Consequently, physicists are experts in everything; other scientists only in those fields of knowledge that are not more basic than their own. But in any case physicists are the best experts (67).

The structures underlying the hierarchical unity of knowledge are derived from mathematics. These structures have to be simple and beautiful. The element of Pythagoreanism inherent in science cannot be overlooked. It is most conspicuous in the work of those physicists who, like Dirac, profess an open commitment to aestheticism in the description of nature; but there are very few exegetes of frontier science who do not stress in one form or another that the laws of physics have to be beautiful in order to be true (68).

Scientific knowledge, mathematically formulated, is out of the reach of the uninitiated. It is a secret knowledge suited to the exercise of power; the ascent of science coincided with the rise of the bourgeoisie, and since then it has remained a tool at the service of the ruling classes all over the world (69).

We now turn to our third precept, the technicality of knowledge. The goal of science — and one of its points of departure — is technical control over nature. The scientific experiment is the prototype of this control.

This control is exercised over phenomena exterior to humanity: the objective world. Humans and their society have to be investigated according to the pattern provided by the hierarchy of knowledge: from biophysics to the formalized social sciences.

The confidence in technical control over nature is absolute and unlimited: this allows the criterion of intersubjective reproducibility to be applied.

On the other hand, questions and experiences which cannot be formulated in terms of technology are not scientific. As a rule, science tends to suppress such questions and to forget them as quickly as possible. This is most clearly demonstrated by the numerous pseudoscientific psycho-physiological theories and practices, from psychoanalysis to parapsychology, to the knowledge contained in the Eastern religions.

But also in social relations a similar process of rejection of non-technical knowledge is under way, which is perhaps less conspicuous because this knowledge has not, or at least not yet, been made the subject of a pseudoscientific discipline. The increasing domination of the world by technology necessitates a deterioration of previously rich and complex social relations.

Social experiences expressed by modes of behaviour, possibilities of social action and reaction (potentials, not norms), fall victim to technical constraint.

Science relegates all these questions to the outside of its own picture of reality. There is no confidence whatsoever in the control of man over himself, and in the self-control of human society. Concepts such as intrasubjective reliability or social reliability cannot exist within science (70).

5.2. *The End of the Myth?*

In the course of its irresistible expansion, science has advanced its frontiers far beyond the domain of natural phenomena accessible to everyday experience. The discovery of remote countries, travels to hitherto unknown shores, amazing inventions have accompanied and promoted the rise of science. The exterior world has been subdued; humans have set their feet on every spot of it. By now, we know everything and the trip to the moon was useless (71).

What next? Shall we throw our despoiled planet on the scrap-heap and move off to new horizons? The idea is not new altogether: it is part of a traditional Utopian scheme depicting with a kind of gloomy enthusiasm the promise of a better (?) future:

Briefly summarized, the radical measures which Bernal prescribed were the following: to defeat the World, the greater part of the human species will leave this planet and go to live in innumerable freely floating colonies scattered through outer space. To defeat the Flesh, humans will learn to replace failing organs with artificial substitutes until we become an intimate symbiosis of brain and machine. To defeat the Devil, we shall first reorganize society along scientific lines, and later learn to exercise conscious intellectual control over our moods and emotional drives, intervening directly in the affective functions of our brains with technical means yet to be discovered. (72)

Undoubtedly, such Utopias are somewhat premature; science has not nearly completed its expansion on earth. Battles are being fought in the hazy frontier land between science and the humanities, in the fields of the social sciences, of anthropology, of ethology, and others more. As Lévy-Leblond (73) points out, frontier science has shifted part of its ideological offensive to these domains. The emotions raised by the debate over the new discipline of sociobiology, for instance, show that there is more at stake than just the resistance against the racist bias of a theory; once again, we are witnessing the 'invasion of a new field' by science.

But the importance of these altercations should not be overestimated. As long as the hierarchy of scientific knowledge is maintained, these frontiers are of secondary importance (74). Now as ever, the extension of physical knowledge to hitherto inaccessible regions of space and time remains the most important section of the frontiers. Here, expansion must continue. A failure of physics, for whatever reason, would have unforeseeable consequences for the self-confidence of science and for its position within society; science would obviously not be all-including; its cognitive hegemony would be questioned.

Yet sooner or later this situation might very well occur in particle physics. Due mainly to the enormous technical difficulties and the financial burdens of experimental research in this field, events allowing a direct interaction between theory and experiment have become increasingly rare during the last 20 years. In spite of colossal efforts, the explosive dynamics of the evolution of early quantum theory, or even of quantum electrodynamics, could not be regenerated. Physics may be reaching a very concrete frontier: a theory of strong interactions, or the great unified gauge theory may simply be too expensive (75).

The guardians of the myth are well aware of this. Hence, the exuberant enthusiasm of the particle physicists when some unexpected phenomenon shows up; a new particle might give rise to talks about 'new physics' — reflecting the desperate hope for a new beginning (76). The exuberance, however, is rarely lasting (77). Hence, too, the redoubling efforts of physicists to stress the basic importance of their work: whatever the experiments yield brings us automatically nearer to an understanding of the world (78). Hence, finally, their conviction that all problems, including those caused by scientific and technological progress itself, can be solved only by still more science and technology (79).

Resignation, however, is spreading. A growing number of scientists begin to doubt whether the world of physical phenomena still has significant new insights to reveal. Some physicists are ready to renounce their avantgarde position in science (80), retaining only the educational background which somehow would still leave them with a leading role (81). Others are looking for radically new frontiers; and, quite akin the pseudoscientists, they fall back on the metaphysical traits which science has carried along ever since Galilei's heroic times, and which have re-emerged with the advent of quantum theory.

We note a return to the inner human self: the vast fields which remain to be discovered there are bright promise for a new colonization which would not be burdened with the original sin of worldly conflicts. Physicists and astronomers are tinkering at a new cosmology which subordinates nature to an ordering spirit (82). Lastly, this spirit is the only reality — the material world just an illusion (83).

This retreat from the material world finally cuts the ground from under the guardians' feet. Already weakened by the difficult economic position it has manoeuvred itself into, frontier science now begins to undermine its ideological strongholds. Aestheticism and technicality, the unequal parents of science, are getting in each other's hair. Their quarrel might foreshadow the decline of the myth of science.

6. Conclusion: Sympathy for the Devil

These dismal prophecies will mean nothing to the average scientist. Even if the gnostics of Princeton had actually succeeded in getting hold of some exceedingly deep wisdom — for the expert standing in the midst of political life, they are just poor lunatics whom the dry air of the campus does no good. The scientists can still be confident: all those activities which could contribute to a truly anti-scientific movement remain on the fringes of society. The cognitive hegemony of science is still almost unchallenged. There is, for the time being, no such thing as an anti-science, which could compete with it.

Nevertheless, the symptoms of dissolution of scientific rationality at the frontiers are not unsignificant; for the hegemony of sciences is being questioned in more than one field. The myth of science is losing ground where it seemed to be strongest: in the domain of economic and political applications — we used to say achievements — of science. The near destruction of humanity in two world wars, the present destruction of the world we live in, the inability to solve the problems of underdevelopment, might be sufficient reasons to shatter the belief that science has still anything to do with progress.

Science, which has so gloriously succeeded in subduing nature, has created a world which it now finds increasingly hard to explain and control. This, at least, is the impression which the layperson gets in view of the experts' inability to realize their ambitious plans, to keep their boastful promises. Other explanations must then take the place of science: no one can live without explanations. If the scientist is unable to predict the effects of the operation of nuclear power plants, someone else will do it.

The struggles at the frontiers of science thus obtain their fierceness from the political struggles which are transforming our present society. No one can say what the new myths will be; the struggles themselves will shape them. Presently, the anti-scientist is all too often compelled to choose whichever ways of knowledge other cultures, which have not yet succumbed to the myth of science, can teach him. Hence, the way inward, and the way to the East. This necessity to rely on pre-scientific myths causes the weakness of an oppositional movement which, when it is politically conscious, certainly does not want to throw out the baby with the bath-water (84).

Just as the scientific myth has incorporated pre-scientific myths, it will itself be recast to contribute to the myths of the future. This is why a political critique of science is indispensable for any anti-scientific endeavour which wants to escape the spell of the magic mirror. This certainly implies a cognitive openness: let us have one, two, many devils, we can't yet know which one will be our next god. But let us not confuse openness and the illusion of an inextant freedom. Be it true, with Feyerabend, that "anything goes", we agree to reply with H.P. Duerr: "You can't fuck them all" (85). Our problem, from now on, must be: whom shall we?

Acknowledgments

We are deeply indebted to the following products of frontier viticulture, which were major sources of inspiration leading us to frontiers far from those of science (86): Valpolicella Superiore, Casa Vinicola Fratelli Bolla, Verona; Vino Chianti, Fattoria Cantuccio, Prop. Contessa Caterina Passerin d'Entrèves et Courmayeur, Tavernelle Val di Pesa, Vend. 1971, D 2710709 and D 2710149; Gumpoldskirchner, Winzergenossenschaft Gumpoldskirchen, Nr. 26009; Sankt Laurent Ausstich, aus dem stiftherrlichen Weingut Trattendorf, Kelleramt Chorherrenstift Klosterneuburg bei Wien, 1975; Meßwein, Stift Klosterneuburg, Kelleramt; Vini del Piave Cabernet, Bianchi Kunkler, Vend. 1975, Nr. 124524 and 142681; Grauvernatsch Alte Reben, 1974; Vino Freisa, Enrico Serafino, Canale d'Alba, 1974; St. Magdalener, Weinkellerei Josef Brigl, Girlan, 1974; and Vino Barbera, del Cantina di Bacco, Casa Vinicola Antonio Vallana & F^{lio}, Maggiora (Piemonte), 1974, Nr. 3777 and 6686, for *very* helpful comments (87).

References and Notes

−1. Any resemblance of the language used in this essay with living English, far from being coincidental, is due to Brigitta Grabner, who was, moreover, indispensable for her knowledge of Shakespeare and her possession of the Oxford Dictionary of Proverbs.
0. Who, in the heroic days of '68, bravely led the first attempts to break out of the monastic conformity of the physics studies at Vienna University.
1. Oswald Wiener, *Die Verbesserung von Mitteleuropa*, roman. Rowohlt, Reinbek, 1969; and further literature quoted therein.
2. How many philosophers of science are there who deal with science as it is done *now*? Even those who allow for discontinuities in their picture of the dynamics of science, like good old Kuhn, must assume that all scientific revolutions are essentially identical. This is again a somewhat ahistoric view. What if the dynamics of science had qualitatively changed in the last 20 or 30 years?
3. D.S. Greenberg, *The Politics of American Science*, Penguin, Harmondsworth, 1969, p. 145.
4. L.C.L. Yuan (ed.), 'Nature of Matter. Purposes of High-Energy Physics', Brookhaven National Laboratory 888 (T−360), 1964.
5. R.P. Shutt, 'Science, Physics, and Particle Physics'. In L.C.L. Yuan (Ed.), *Elementary Particles: Science, Technology, and Society*, Academic Press, New York and London, 1971, p. 9.
6. *Ibid.*, p. 43.
7. V.F. Weisskopf, 'The Frontiers and Limits of Science', *American Scientist* 65, 405 (1977).
8. P.M. Morse, 'Is Physics too Ingrown?', *Physics Today*, April 1973, p. 23.
9. M. Bunge, *Scientific Research, I: The Search for System*, Springer-Verlag, Berlin, Heidelberg, New York, 1967, p. 32.
10. *Ibid.*, p. 33.
11. W. A. Fowler, 'Physics in 1976 − A Personal Account', *Physics Today*, April 1977, p. 41.
12. M. Bunge, *Scientific Research, I, loc. cit.*, p. 34.
13. V. F. Weisskopf, 'The Frontiers and Limits of Science', *Loc. cit.*
14. R.P. Shutt, 'Science, Physics, and Particle Physics', *loc. cit.*, p. 31.
15. F. J. Dyson, 'The World, the Flesh, and the Devil'. In C. Sagan (Ed.) *Communication with Extraterrestrial Intelligence*, MIT Press, Cambridge and London, 1973, p. 371.
16. M. Gell-Mann, 'How Scientists can really help', *Physics Today*, May 1971, p. 23.
17. H.L. Davis, 'Editorial', *Physics Today*, March 1971.
18. "The Devil always leaves a stink behind him". *The Oxford Dictionary of Proverbs*, Oxford University Press, 1975.
19. K.M. Michel, 'Schön sinnlich', Kursbuch 49, Kursbuch-Verlag, Berlin, 1977, p. 11.
20. M. Bunge, *Scientific Research, I, loc. cit.*, p. 36.
21. K.M. Michel, 'Schön sinnlich', *loc. cit.*, p. 12.
22. M. Bunge, *Scientific Research, I, loc. cit.*, p. 36.
23. M. Gardner, *Fads and Fallacies in the Name of Science*, Dover Publications, New York, 1957.

24. M. Bunge, *Scientific Research, I, loc. cit.*, p. 39.
25. We are well aware of the desperate shrug of the Feyerabendians' shoulders at these harsh words. Just skip this section if it bores you.
26. M. Bunge, *Scientific Research, I, loc. cit.*, p. 256.
27. F.J. Dyson, 'The Future of Physics', *Physics Today*, September 1970, p. 23.
28. The subscribers of magazines like *Science, Physics Today* or *la Recherche* carry out heated discussions of this subject in the 'Letters to the Editor'. One consequence of these public debates is that e.g. parapsychologists have to expose planning, realisation and interpretation of their experiments to an extent unthinkable to the majority of the rank and file scientists.
29. !
30. O.M. Bilaniuk and E.C.G. Sudarshan, 'Particles Beyond the Light Barrier', *Physics Today*, May 1969, p. 43.
31. P.A.M. Dirac, 'The Evolution of the Physicists' Picture of Nature', *Scientific American* 208, May 1963, p. 45.
32. S. Weinberg, 'The Future of Unified Gauge Theories', *Physics Today*, April 1977, p. 42.
33. O.M. Bilaniuk and E.C.G. Sudarshan, *loc. cit.*
34. M. Bunge (Ed.), *Quantum Theory and Reality*, Springer-Verlag, Berlin, Heidelberg, New York, 1967.
    J.M. Jauch, *Are Quanta Real? A Galilean Dialogue*, Indiana University Press, Bloomington, 1973.
    J.M. Lévy-Leblond, 'Towards a proper Quantum Theory', *Dialectica* 30, 161 (1976).
35. C.M. Patton and J.A. Wheeler, 'Is Physics Legislated by Cosmogony?'. In C.J. Isham, R. Penrose and D.W. Sciama (Eds.), *Quantum Gravity*, Clarendon Press, Oxford, 1975, p. 538;
    To which M. Bunge would reply:
    > If ... we claim that a given theory concerns not only microsystems but also ourselves, then we must justify this claim by exhibiting all the variables and the formulas that demonstrably point to ourselves. If we are serious about it we shall have to bring in the whole of psychology: we shall thus establish the science of quantum psychology ... But, whatever way we choose, let us be clear what we are talking about: otherwise our very sanity will be questioned. (M. Bunge, 'A Philosophical Obstacle to the Rise of New Theories in Microphysics'. In T. Bastin (Ed.), *Quantum Theory and Beyond*, Cambridge University Press, Cambridge, 1971, p. 263).

    What a mess! On which side of the frontier are we just now?
36. Y. Nambu, 'The Confinement of Quarks', *Scientific American*, November 1976, p. 48.
37. "Basic statements must be testable, intersubjectively, by 'observation' ", says our much-beloved Sir Karl Raimund (*The Logic of Scientific Discovery*, Hutchinson, London, 1968, p. 102).
38. This is what many first-year physics students learn:
    > Incidentally, psychoanalysis is not a science: it is at best a medical process, and perhaps even more like witch-doctoring. It has a theory as to what causes diseases — lots of different 'spirits', etc. ... (The *Feynman Lectures on Physics*, Vol. 1., Addison-Wesley, 1963, pp. 3–8.)

39. C.M. Patton and J.A. Wheeler, 'Is Physics legislated by Cosmogony?', *loc. cit.*
40. "Actually, contemporary science seems to have succeeded, in a bold backward move over millions of centuries, to make itself the witness of that initial Fiat Lux. . . . ." (Pope Pius XII in an address to the Pontifical Academy, 1951; cited in H. Alfvén, 'La Cosmologie, Mythe ou Science?' *La Recherche* **69**, 610 (1976).)
41. C. Sagan (Ed.), *Communication with Extraterrestrial Intelligence, loc. cit.*
42. G.K. O'Neill, 'The Colonization of Space', *Physics Today*, September 1974, p. 32; and a flood of secondary literature which has been published during the last few years.
43. T.B.H. Kuiper and M. Morris, 'Searching for Extraterrestrial Civilizations', *Science* **196**, 616 (1977).

    Decidedly, colonialism has profoundly affected the scientists' thinking; the idea of discrete watching, however, is a little more recent, reminding one of the Cold War and the like. By the way, this hypothesis has the same logical status as that of quark confinement. Since "there can be little doubt that civilizations more advanced than the earth's exist elsewhere in the universe" (C. Sagan and F. Drake, 'The Search for Extraterrestrial Intelligence', *Scientific American*, May 2975, p. 80), it must be explained why we are not regularly holding joint conferences with them.
44. J. Lederberg, 'Searching for GODot'. In C. Sagan, *Communication with Extraterrestrial Intelligence, loc. cit.*, p. 393.

    Boy: What am I to say to Mr. Godot, sir?
    Vladimir: Tell him ... *he hesitates* ... tell him you saw us. *Pause*. You did see us, didn't you?
    Boy: Yes, Sir.
    He steps back, hesitates, turns and exits running.
    (S. Beckett, *Waiting for Godot*, Suhrkamp, Frankfurt, 1971, p. 132).
45. T.B.H. Kuiper and M. Morris, *Searching for Extraterrestrial Civilizations, loc. cit.*
46. J.A. Hynek from the Center for UFO Studies, Evanston, in a letter to *Physics Today*, December 1976, p. 46.
47. P. Feyerabend, *Against Method*, London 1975.
48. This is not only the case for those pseudosciences which, like parapsychology, are only exceptionally admitted as part of an academic institution, but also for those disciplines of the social sciences which are not prepared to acknowledge the hegemony of science. *Cf.* the recent discussion about the qualification of social anthropologists who applied for funds at the U.S. National Science Foundation (Science, 25.2.1977, and 22.4.1977). The NSF representative in charge was deeply scandalized:

    Authors of some proposals would write that they wanted to study a particular group simply because it is disappearing or no one had ever described it before.
    Considering that a large part of current scientific work is devoted precisely to the study of effects and phenomena whose existence is known, but which have never been described before, this accusation seems peculiar . . . Not to mention the study of groups which may not exist at all — such as extraterrestrial ones.
49. R. Collins, 'Toward a Modern Science of the Occult', *Consciousness and Culture* **1**, 43 (1977).
50. R. Chauvin, 'La Parapsychologie — après le contre, le pour', *La Recherche* **69**, 659 (1976).

51. L. Oteri (Ed.), 'Quantum Physics and Parapsychology', Proceedings of an International Conference held in Geneva, Switzerland, August 26–27, 1974. Parapsychological Foundation, Inc., New York, 1975.
    This conference was attended almost exclusively by partly well-known physicists who wanted to contribute to the explanation of parapsychological phenomena. G. Feinberg, a theoretical physicist known for his book *The Prometheus Project*, spoke at length about 'Precognition – a Memory of Things Future' (made possible by the advanced solution of Maxwell's equations ...), and brought some very recent results from quantum chromodynamics into the debate in order to question the apparent incompatibility of parapsychological effects with the 'rest of physics' (p. 248). The physicist Costa de Beauregard, on the other hand, affirms:
    > Relativistic quantum mechanics is a conceptual scheme where phenomena such as psychokinesis or telepathy, far from being irrational, should, on the contrary, be expected as very *rational*. My thesis is that they are postulated by the very symmetries of the mathematical formalism ... (p. 101).

    Again anything goes, provided we let mathematics do it and physicists spell it out!
52. J. M. Lévy-Leblond, 'Mais ta physique?'. In: Hilary Rose and Steven Rose (Eds), *L'idéologie de/dans la science*, Editions du Seuil, Paris 1977, p. 112.
53. P. Jordan 'Die weltanschauliche Bedeutung der modernen Physik' Schriftenreihe der Liga Europa, Klinger-Verlag, München, 1971.
54. It follows that those analyses trying to explain pseudoscience as a replacement for religion miss the actual problem: science itself can serve very well as religion.
55. H. Alfvén, 'La Cosmologie – Mythe ou Science?', *loc. cit.*
56. M. Gardner, *Fads and Fallacies in the Name of Science, loc. cit.*, p. 68.
57. Should the Acropolis never have been built because it was a waste of money? (R.P. Shutt, 'Science, Physics, and Particle Physics', *loc. cit.*, p. 3).
    > (Basic scientific research) bears much the same relation to contemporary civilization that the great artistic and philosophical creations of the Greeks did to theirs, or the great cathedrals did to medieval Europe. (H. Brooks: Future needs for the support of basic research, cited in: P. Weingart: Die Amerikanische Wissenschaftslobby. Bertelsmann-Universitätsverlag, Düsseldorf, 1970, p. 53).
58. Major General G.J. Keegan, former head of U.S. Air Force Intelligence, is convinced that the Soviet Union is using proton beams from a particle accelerator as an antiballistic missile weapon. Physicists are mostly very sceptical about this assertion, the technical problems involved being enormous. This does not deter the General, who accuses the scientists of their allegiance to conventional wisdom, and of believing that the Russians could not succeed where the Americans had failed. Actually, a U.S. project named 'Seesaw' aimed at developing an ABM electron beam weapon was closed down in 1973. Still, the U.S. are spending 7.5 million dollars a year on beam weapons. *Cf. Science* **196**, 407, 957 (1977).
59. A.W. Saénz *et al.*, 'Telecommunication with Neutrino Beams', *Science* **198**, 295 (1977).
60. An inventory of the technologies originating from research in particle physics includes the following items: very high-voltage techniques, e.g. transmission lines for 1 MV or more, or applications for the spraying of paints; high-frequency power tubes and their by-product, electronic cooking; supervacuum technology, and

vacuum processing of foods and juices; devices for pattern recognition, of possible future use in the keeping of records in banks and post offices; superconducting devices. (J.P. Blewett, 'Interactions between elementary particle research and engineering'. In L.C.L. Yuan (Ed.), *Elementary particles: Science, Technology, and Society, loc. cit.*, p. 290 ff.).
>Technological spinoff, however, is a topic which can make engineers and solid-state physicists somewhat apoplectic. They emphatically do not tend to feel that it is a valid way of advancing technology to subtract X million dollars from their own budget and spend some of it on a superconducting linac rather than to spend it directly on say superconducting generators and power lines...
(P. Anderson, 'Are the big machines necessary?', *New Scientist*, September 2, 1971, p. 510).

61. G.K. O'Neill, the leading prophet who would like to make us believe that life in space colonies is not only technically feasible but also exceedingly charming, has won his spurs, too, in accelerator design: he was one of the first to develop the idea of particle storage rings. (*cf. New Scientist* 74, 718 (1977)).
62. S. Weinberg in an interview in *New Scientist* 73, 404 (1977).
63. *Ibid.*
64. Unless otherwise indicated, the quotes in this section are taken from the brochure: 'The War Physicists: documents about the European protest against the physicists working for the American Military through the Jason Division of the Institute for Defense Analysis', edited by B. Vitale, University of Naples, 1976.
65. "What does Weinberg do to relax? He reads history; he has a fascination for medieval England . . . ; and would prefer above all to have been a poet . . . In the slack period between 1967 and 1971 he did military work on strategic weapons . . ." (*New Scientist* 73, 404 (1977)).
66. H. York, 'Eisenhower's second warning', *Physics Today*, January 1977, p. 9.
67. Actually one had only to look at the Jason membership lists to reach this conclusion.
68. "All important physicists of the scientific age were filled with this Pythagorean belief in the rationality and the mathematical beauty of the structure of our cosmos . . . I would call this Pythagorean-metaphysical belief the mystical element of the exact sciences." (V. Gorgé, 'Das mystische Element in den Naturwissenschaften'. In A. Mercier (Ed.), *Mystik und Wissenschaftlichkeit*, Verlag Herbert Lang, Bern, 1972, p. 107).
69. Should we expect the philosophers of science to be free of a little bit of class conscience?
>By virtue of its spiritual power and its material fruits science has become to occupy the center of modern culture – which is not to say the culture of our days. In fact, it would be foolish to forget that, alongside higher culture, folk culture still lingers. And pseudoscience occupies in contemporary urban folk culture a position similar to the one held by science in higher culture. (M. Bunge, *Scientific Research, I, loc. cit.*, p. 34).
70. Knowledge gained in personal or social experiences necessarily gets lost. Such experiences do not count as experiments. Accordingly, they are discouraged or even prohibited. Other civilizations seemed to make a better use of such possibilities:
>Whatever be the goal at which they aim, the most striking peculiarity of Tibetan

mystics is their boldness and a singular impatient desire to measure their strength against spiritual obstacles or occult foes. They seem animated by the spirit of adventure and, if I may use the term, I should call them 'spiritual sportsmen'. (A. David-Néel, *Initiations and Initiates in Tibet*, Rider and Company, London, 1970, p. 15.)

71. Folk culture, indeed, has something relevant to say about higher culture; it has a theory of it, from which it makes useful predictions. This is shown, for instance, by this German popular tune of the late fifties:
    Die Fahrt zum Mond hat sich gelohnt.
    Jetzt weiß die Wissenschaft —
    im Grunde ganz gewissenhaft —
    daß sich die Fahrt zum Mond nicht lohnt.
    Drum hat die Fahrt zum Mond
    im Grund sich doch gelohnt.
72. F.J. Dyson, *The World, the Flesh, and the Devil, loc. cit.*, p. 373.
73. J.M. Lévy-Leblond, 'Mais ta physique?', *loc. cit.*
74. Actually, there are timid manifestations of protest within the hierarchy. Chemistry, biology, etc., can be reduced to physics, but only in principle, the necessary calculations being far too complicated in most applications. Does this not mean that each of the natural sciences has in fact its own autonomous method adapted to the problem it studies?
    ... The objects, concepts, and ideas that the scientist uses when he tries to understand what goes on do not deal directly with atoms but rather with the structures that are immediately involved in the phenomena under study. (V.F. Weisskopf, 'The frontiers and Limits of Science', *loc. cit.*).
    This is the concession even the physicist must make. With the advent of tools like systems theory, hierarchical ways of thinking lose their importance. The consequences of this evolution for the unity of science cannot yet be foreseen.
75. "... I feel that the long-range prospects for going on with accelerator physics in this style are not good. Even if the most optimistic assumptions about the Batavia machine are realized and we find an exciting new world of phenomena in the hundred GeV range, we are nevertheless running into a law of steeply diminishing returns." (F.J. Dyson, 'The Future of Physics', *loc. cit*).
76. 'Gauge theory predictions and muon decay', *CERN-Courier* **17**, 51 (1977).
77. A recent report about progress in particle physics is entitled: 'Where are we now?' It reads: "Perhaps the simplest and most honest answer to the question in the title is that we do not know." (*CERN-Courier* **17**, 407 (1977)).
78. The same report says further on: "However, we have certainly moved nearer to understanding the workings of Nature as a result of the research in 1977".
79. "Suppose we are now indeed to abandon the old principle of building anything we know how to build because we know that it can be done ... Suppose we now try to include also the cost in human, social, and environmental terms and to deploy, as I believe we should in the future, only a smaller and smaller fraction of what is technically possible, according to the principles of technology assessment and control. This is what I call the 'narrowing cone'.
    If indeed we make use of a narrowing cone of the technical possibilities generated by science and engineering, does that mean that we need less technology and

less applied science? I claim it means we need a good deal more technology and more applied science, and in the long run, more pure science as well." (M. Gell-Mann, 'How Scientists Can Really Help', *loc. cit.*)

80. "I take it as self-evident that physics will not flourish in isolation from the rest of science. In particular, physics should keep in close touch with biology, as biology rather than physics is likely to be the central ground of scientific advance during the remainder of our century." (F.J. Dyson, 'The Future of Physics', *loc. cit.*)

81. "An individual physicist, working in close collaboration with engineers and chemists and biologists, may well be able to make important contributions. However, he should not expect that what he does in the environmental field will be mainly physics. If he is any good, he will use his physics only as a cultural background in thinking about problems that are primarily chemical, biological or political in nature." (*Ibid.*).

82. R. Ruyer, *La Gnose de Princeton*, Paris, 1976.

83. F. Capra, *The Tao of Physics*, Wildwood House, London, 1975.

In this book, which was extraordinarily well received by the reviewers of *Science* and *Physics Today*, the author tries to show that recent developments in particle physics — namely the 'bootstrap approach' — indicate a flawless unity of the basic patterns of nature with the thought patterns of Eastern religions and philosophies. Thus science itself proves that the world which it has helped to shape has been constructed according to a wrong ground plan; after several centuries of excellent sales, technological society is shown to be a waste product. Will it be able to pull itself out of the swamp by its own bootstraps, as did old Münchhausen? There is unfortunately no answer to this question.

To those who are afraid of the mathematical intricacy of physics, Capra has a warm comfort to offer:

> ... The following discussion may appear to be rather dry and technical. It should perhaps be taken as a 'yogic' exercise which — like many exercises in the spiritual training of the Eastern traditions — may not be much fun, but may lead to a profound and beautiful insight into the essential nature of things. (p. 135)

An insight which could be as follows:

> ... The possibility that the hadron patterns will some day be derived from the general principles ... must be taken seriously ... If this turns out to be true, modern physics will have come a long way towards agreeing with the Eastern sages that the structures of the physical world are *maya*, or 'mind only'. (p. 293)

84. Actually anti-scientific movements are all too often an easy prey to reactionary political forces. It may be funny when Feyerabend asks why one has not the possibility, at school, to study both evolution theory and creation theory in the biology class. It is much less fun when the same claim is raised by the ultra-conservative fundamentalist churches of the southern states in the U.S.A. (*Cf.* D. Nelkin, 'The Science Textbook Controversy', *Scientific American* **234**, October 1976, p. 33.)

85. H.P. Duerr, 'Können Hexen fliegen?'. In *Unter dem Pflaster liegt der Strand*, Vol. 3, Berlin 1976, p. 75.

This methodological principle can be traced back to Malinowski — we hope to be justified in our assumption that we, honouring our scientific forefathers as we

do, can return to the same source the accusation of male chauvinism that is sure to be brought forward against us.
86. 'There is a Devil in every berry of the grape.' The Oxford Dictionary of Proverbs, *loc. cit.*
87. Oh thou invisible spirit of wine,
   if thou hast no name to be known by,
   let us call thee devil.

(Shakespeare, The Tragedy of Othello, Act Two, Scene Two).

ALTERNATIVES IN SCIENCE – ALTERNATIVES TO SCIENCE?

GERNOT BÖHME
Technische Hochschule, Darmstadt

This essay is a retrospective reflection on the project 'Alternatives in Science' carried out at the Max-Planck Institute in Starnberg for the Study of the Conditions of Human Life in the Modern World. The aim of the project was to examine the determination of alternatives in science through socially determined problem-situations and goal-orientations. The purpose of the present essay is to show in what way the possibility of such alternatives could be thought of within the frame of the epistemology and philosophy of science. Tracing out this possibility constitutes the basis for the main thesis: *There exists a social interest not only in the utilization of scientific knowledge but also in its production.*

1. Reasons for Requesting Alternative in Science

The background of the question concerning alternatives in science is the present critique of science. Of course, the development of science has always been accompanied by criticism (1) and the current criticisms largely only raise long-familiar topics: that scientific knowledge is abstract, one-sided, esoteric, not adequately grounded. Yet there obtains today a particular unison in the various facets of this critique in that it is directed against what from Bacon onward has constituted the social legitimation of the new learning: its significance in providing enlightenment and the expectation that it would serve to improve the conditions of human life. The task of providing enlightenment appears to have been fulfilled and thus the cultural value science represents – where it has not become perverted into a 'scientistic world-view' – has been called into question. Further, the manifest ambivalence of scientific-technical progress has eroded the belief that scientific progress is in all cases identical with human progress.

The thrust of the present critique of science is aimed against its social isolation (scientific communities, the ivory tower), the gulf between the scientific expert and the layperson, the internal hierarchic form of the organization of science, etc. Its determinant is the question: Whom does science serve? (Bernal).

On the whole, this critique does not appear to question the cognitive state of science. It is a critique which might be understood in the sense that if the social integration of science were different, if the chasm between scientists and the laity were closed, if it were possible to rationalize the advance of science by conscious research planning to take account of the needs of society, then the ambivalence inherent in present-day science could be eliminated. But the critique goes further than this. The traditional philosophical critique that scientific knowledge is not justified (2) has turned into the reproach that science is blind to what it does ("science does not think", Heidegger). The romantic criticism of the limitations of modern natural science is now intensified in the characterization of contemporary scientific knowledge as *Herrschaftswissen* (3) (4), that is, knowledge for mastery, domination, control. The ambivalence inherent in scientific-technological knowledge is identified as the ambivalence of a purely instrumental knowledge, as the reduction of reason (Vernunft) to understanding (Verstand) (Horkheimer). It is argued that the ambiguity of knowledge is rooted in its concept of truth which is that of a factualist truth, and in the linearity of causal thought. It is held that the destructive tendencies are based in the method itself, the method of idealization, abstraction, and quantification: in the de-qualification of nature.

This form of criticism gives rise to the demand for not only a different social organization of science but for 'a different sort of science'.

This means a science which is directly and not only incidentally linked to human needs, a natural science which treats nature not only exploitatively but from the start avoids any ambivalence as to its applications. If this demand is not to be left to fate in a diffusion of desiderata and if it is not to be subjected to ideological distortions, then it must be made explicit that this demand implies questions addressed *to* modern science, that is questions which ultimately can only be answered on the basis of modern science.

2. The Context of Reception

The question concerning alternatives in science itself meets with resistance and is likely to be distorted in public discussions — from the side of Anti-Science Movement, of institutionalized Marxism as well as the side of conservative science ideology. It is the last type of resistance which will be discussed more in detail in the present essay because it is within this setting that our theses about the alternatives in science have aroused a politically inspired polemic.

2.1. Anti-Science Movement

In the Anti-Science movement (5) the criticism of existing science becomes a criticism of science as such. Alternatives are not sought in science but rather in meditative forms of cognizing, respectively, in immediately-communicative interaction with nature. In as far as 'a different kind of science' enters the discussion at all, the concept entails romantic and ambiguous notions which either presuppose or postulate the direct unity of humanity and nature. The rehabilitation of other forms of cognition, the critique of the monopoly position occupied by modern science turns here into a tendency to abolish science. The path Aldous Huxley traversed from his critique of the new scientific-technological civilization to psychedelic practice, seems paradigmatic for this turn of mind (6).

2.2. Institutionalized Marxism

Where Marxism as Marxism-Leninism has become the official world-view of the state, or among the theorists who are its proponents, there prevails a scientistic and technocratic understanding of science: An epistemology informed by Lenin's conception of knowledge as the reflexion of objective reality in man's consciousness, simply rules out the question concerning socially determined alternatives in science. Here, scientific realism is bound up with the belief that science is an instrument designed to promote social progress. The experience of particular branches of modern science having been cast out and rejected by the ideology and the effects of this have at present produced an emphasis on the system-neutrality of science, the assertion that the problem of 'good science' depends solely on the social integration of science (7). The maxim of 'electrification and socialism' may stem from the science-rooted optimism of the nineteenth century, but even today the faith in science as an instrument is relatively intact. According to the theory of the scientific-technical revolution, science becomes the major productive force. Ultimately, its development constitutes the base for the transition to communism (8). Viewing science, in particular natural science, as problematic or relativizing it in respect to its social context is described as social defeatism (9).

2.3. *The Conservative Ideology of Science*

The conservative ideologists of science sidestep any substantive discussion of alternatives in science because of the social interest attributed to the production of science. This social interest is perceived as a demand for a social legitimation of science and on the other side the claim of society to control science. Thus Albert (10) writes: "According to this reading, science which, as a branch of social division of labor, is a social domain, is required in each case to legitimize itself by trying to demonstrate the practical relevance of its problems and results to social needs." And, further, the claim "that there exists a connection between knowledge and practical needs which is decisive for the structure of science" makes it seem plausible "to subject even basic research to external control, which would be fatal for scientific progress" (11)

The arguments by which the conservative ideology of science averts the question about what social alternatives are possible within science are the following: The linkage made between alternatives in science and social interests breaks down the separation between facts and values. Further, values are subject to social, respectively political decisions: "The . . . separation between political legitimacy achieved on the basis of the procedures by which norms are established and scientifically grounded assumptions about the truth content of hypotheses arrived at by virtue of empirical testing procedures is abandoned" (12). In this way, so the argument runs, what is taken to be scientifically true comes to depend on political decisions, which means that science becomes impossible. Poser (1976) has summed up this line of argument, advocated primarily by Albert and Radnitzky, by speaking of "the subordination of the truth concept to the primacy of politics". In his summary of the discussion he said "However, when external political or social authorities decide the question of what is true, the sciences are deprived of the methods of testing constitutive of the scientific procedure"(13).

To parry the demands for social legitimation and the claims to social control, the conservative ideology of science thus stylized the question concerning socially determined alternatives in science into a different question, one which leads to pseudo-science. These arguments are rooted in a concept of rationality whose progress always depends on differentiation (14). Any attempt to overcome the analytical separations, in this instance the separation between fact and alue and between knowledge and human interest, are viewed as regression. A premium is given to understanding in opposition to reason, and the plea is for an engineering type of expertise and proficiency to counter the 'domination' of the intellectuals (15), (16).

3. What is Science?

If one systematically averts the question of alternatives in science, interpreting it as constituting alternatives *to* science, it may be supposed that one knows what science is. In fact, this opposition is but one tendency in the historical controversy about the definition of the nature of science. The various attempts made by philosophy of science to define the nature of science ahistorically, that is to draw a demarcation between science and non-science, can by now only be considered to have failed. Not the attempts made in linguistic analysis (criteria of meaning), nor methodological approaches (verifiability, falsifiability) nor those in line with an ethics of science (criticism) have produced satisfactory results. In retrospect all these are also but attempts to reach a definition of science by rejecting other cognitive developments, as for instance recently those of psychoanalysis and Marxism (17).

Demarcation is an historical process. What we recognize to be science today is the product of the rejection of other intellectual enterprises, from astrology and alchemy, through Mesmerism and Naturphilosophie to acupuncture. It is one of the most promising approaches of a materialistic history of science to describe the formation of science as a social process, as the Social Construction of Scientific Knowledge (18). This happened through the quarrels between groups supporting alternative types of knowledge — the hierarchization of types of knowledge being the consequence — or through some sort of self-purification of one intellectual circle by which it managed to institutionalize itself and so to exclude other types of knowledge as pseudo-science, literature, etc. So, for example, it can be shown that modern science managed to become 'positive science' only by reducing its program, leaving parts of the original endeavours outside science.

In his study of the social construction of science W. van den Daele (19) has shown that in the seventeenth century the new learning was associated with political, social reform, and educational aims. And further that it was the political conditions arising form the Restoration which made for a situation in which the institutionalization of science in such scientific associations as was the Royal Society or the Académie Royale was only possible on condition that such far-reaching claims were given up. The consequence was that for centuries to come politics, education and society were screened out of the program of science.

Even though science has arrived at a self-definition by rejecting or disregarding other intellectual enterprises, there is little hope, as a sort of countermove, to revalue these other intellectual projects as 'sciences' and seek alternatives within this context of thinking. For, whatever the truth-content of the 'pseudo-sciences' may be, the fact that they were rejected did not allow them to develop into forms of knowledge which admit of generalization. A more promising approach is to make clear the bounds of the cognitive venture that is modern science, this on the basis of a study of those sciences which were able to advance outside the area dominated by modern science. True, there are but a few examples of this type, such as the science of antiquity, science in China (20), or Goethe's theory of colours.

According to Plato, knowledge only becomes science when it contains a complete survey over all possible objects within a particular domain and reveals the essential systematic connections between them. In this sense science can be completed as to its content, and is not, as modern science is held to be, an infinite progression of knowledge. The procedure of hypothetic deductions in these terms is not, as in the case of modern science the real virtue of science but rather a state resulting from science's lack of a foundation, and thus a state which must be overcome. The content of science then is not the lawlike connection between events but the systematic of the objects of knowledge. The practical import of science thus lies not in the possibility productively to change the world but in the faculty to orientate oneself within the existing orders of objects and events (21).

Goethe's interest in knowledge focussed on the sensory significance (the image-effect) of colors, and not, as was the case with Newton, on the construction of optic apparatus.

For this reason, for Goethe the last instance of verification was the perception of the senses while for Newton as a representative of modern science it was the effect obtained in and by the apparatus. The theme of Goethe's theory of colors is the condition in which colors appear. Newton's theme were the components and properties of light. According to Goethe there exist objective and subjective conditions for the appearance of colors. Newton, by contrast, sought to determine the objective facts and then to connect these with their effects in perception by virtue of a psychophysical law. Goethe's aim was to know the inner order of the phenomena, for Newton it was the isolation and explication of effects from dispositions of light (22).

If we compare present-day science with non-contemporary types of scientific inquiry this will show the limitations of present-day science, its addition to Cartesian dualism, its elementarism, and its disregard for the phenomenal realities, for existing orders. This comparison evinces the conceivability of, in principle, even radical attempts to form alternative sciences. But conceptions alone are not of much use, it is the 'doing' that is decisive. On the other side, it is worthwhile for philosophy of science and epistemology to reflect on how the potential of modern science might be developed so that it could regain perceptibility for excluded domains of reality and a relationship to the good.

Moreover, there are pragmatic arguments for not trying to seek alternatives outside the framework of modern science: Modern science has been able largely to assimilate the positions of its rivals, at least in the long-term. In doing so it has evidently itself changed. A study of the relationship of modern natural science and German natural philosophy (Naturphilosophie) exemplifies this clearly. The assimilation of the positions held by natural philosophy was bound up with the overcoming of the mechanistic line of thought in physics and the development of the field concept. The positions of vitalism were assimilated in a similar way. Even Liebig still believed that a particular vital force must be presupposed which bears on the formation of higher orders of matter. Today, one can account for this force chemically, but by means of a chemistry which has itself changed by integrating the concept of information (molecular genetics). By analogy, mechanics as global analysis is setting out to conquer mechanical complexes, such as the planetary system, the explication of the order of which, Newton for instance, still left to theology. René Thom's theory of catastrophy has developed means by which discontinuities in nature can be precisely grasped, the transformation of 'quantity into quality' can be made intelligible, a question which from the vantage of modern science had so far been left to an obscure dialectic. It may be added that alternative approaches in science which were suppressed on political reasons are often reintegrated later when the climate has changed. An example for this is Virchows' 'political medicine', a project from 1848, which is nowadays well institutionalized under the heading of social medicine (22a). This case reminds us not to forget the 'costs' of the suppression of alternative types of knowledge and science when emphasizing the flexibility of modern science 'in the long run'. These costs may be social ones when certain issues are not to be thematized as scientific problems in the meantime (conditions of working and housing in the case mentioned) and they may lie on a cognitive level as well (colours being excluded as chemical issue by Lavoisier's chemistry, an example of Kuhn's) (23).

Yet the historical flexibility of science is reason enough to proceed from the existing fund of knowledge when seeking for alternatives today. An additional argument is the necessity of maintaining this fund of knowledge. Evidently, to try to give universal validity to a systematic cognitive venture, even if it is in its own right successful, is useless, if it is not linked to the advancement of modern science. The fate of Goethe's theory of colours, the scientific nature of which and its practical justification as a 'science proceeding by sense-perception' can easily be seen today, is a warning example. That the theory of colours did receive some recognition at least as a point of controversy was due solely to the stature of its author and to his reputation in another field of intellectual creativity. The conclusion is therefore: Even if what we hold to be science is to be viewed as the product of an historical and often violent process of demarcation the question concerning alternatives, motivated by critique of this science, must ask for alternatives *within* science.

4. Alternatives in Modern Science

This search for alternatives motivated by science criticism is concerned with alternatives of varying social significance, alternatives which depend on various social problems, various value- and goal-orientations. Our initial approach (24) was that the different social relevance of sciences was a result of the different interest the ruling fractions of society put in the cognitive alternatives science 'neutrally' provided.

For this reason, we distinguished between two types of development, one in which the evaluation of science proceeds naturally, and one in which it proceeds intentionally. In respect of the former the social significance of a theory would be reflected in a quasi-Darwinistic way in the chances for its advancement provided by society, in the latter it would manifest itself in the conscious planning of particular areas of research.

But it was clear from the first that the different premiums accorded by society to various alternatives in science had to find their base already in the cognitive content of the respective project. And it was also clear that the possibility of intentionally advancing science according to social purposes has its precondition within the state of a science itself. This led to the thesis that science is related to social interest already on the level of its content (and consequently is not just to be applied socially as a finished instrument of knowledge in this or another way). I therefore will attempt to explain the possibility of the linkage between the content of a science and social goals by two theoretical concepts: the concept of the 'constitution theory' which examines the social connection of science for the case of the emergence of new disciplines, and the concept of 'finalization' which traces out the social connection in already developed sciences.

4.1. Constitution Theory

The object of a science is not to be understood as an object with definite properties, which simply exists in nature irrespective of the way people scientifically thematize it. Rather, *what* the object of a science is will be determined by the way it is cognized. One should not, however, make the error of asserting that one can point to substantive elements of nature, that is 'the qualitative', particular phenomena, effects or forces, as the extreme constructivism of, for instance, the Dinglerian type had claimed (25). It is only the *formal* properties of the object, and thus not the phenomena, but the object's phenomenality (26) that are determined by the subject of knowledge. The formal properties of the objects of scientific research are informed by the standardization of scientific experience and the development of specific forms of scientific discourse and of scientific language. Thus, for instance, the type of quantity to which a particular phenomenon belongs viewed scientifically is established by virtue of the principles according to which it is quantified. These principles are the rules according to which the respective phenomenon is operatively and technically investigated by science (27).

The standardization of the empirical cognizing of objects which alone makes experience scientific dissolves the pre-scientific, diffusely given unity of the object into multiple data. And from this manifoldedness of data the unity of the object is reconstructed by means of scientific laws and of theory building. But the fact that there exist lawlike regularities in nature and that the manifoldedness of the phenomena is organized into objects is not what is derived from the subjective element of cognition. Rather the product of subjective cognizing is the form of the laws (lawlikeliness) and the concepts by which objects are cognized (the theoretical unity of experience). Lawlikeliness and the concepts denoting scientific objects are engendered by the standardization of discourse. By these means it is established, for instance, what is to be regarded as natural and what is a phenomenon which requires an explanation (28), what is to be regarded a cause and what an effect, how object and properties of the object are to be distinguished from each other.

The rules according to which one makes something the theme of a scientific study depend in part on the manner in which such a something has become one's problem. Thus, the interest one has in an object may have consequences for its formal properties. In this sense, the sociologists of knowledge, from Scheler to Habermas, have asserted that for modern natural science nature only becomes an object of interest to the extent that it becomes interesting to manipulate it technically. To be sure, such statements have only been made in a global sense, and have become more specific only in relation to methodology. They have an effect on the object itself only by the pragmatistic claim that dispositional predicates are action-dependent. Yet there are means by which one can show in a much more specific way how the knowledge-interest determines the rules of the empirical approach to the object and thereby what must be regarded as 'given' from the scientific point of view. The comparison between Goethe's and Newton's theories of colour is an example of this kind. Here, the alternative interests, via alternative experimental rules, leads to different conceptions of what, in principle, is to be regarded as the 'phenomenon of colour'. Perhaps more impressive, because it is a part of the modern history of science, is the controversy between Titchener and Baldwin which I have analyzed as a discussion between alternative thematizations of a particular psychic phenomenon (29).

The lifeword phenomenon which is the topic of this controversy is the so-called personal equation, that is the fact that there exists an individually determined delay between stimulus and conscious perception of the stimulus (reaction time) (30).

Performing reaction-time measurements it had been possible to distinguish two classes for reactions separated by a characteristic time difference. In measuring this difference diverse data were produced by the Leipzig School and Baldwin. My analysis has made explicit the background of this discrepancy. The knowledge interest of the Leipzig school was to establish psychology as natural science proceeding by experiment. So they conceived of the sensory/muscular difference as a phenomenon which, as such, had to be produced as pure as possible by the experiment, and which had to be explained on the basis of the assumption of internal psychic apparatus being equal for all persons. The Leipzig School therefore selected subjects for their experiments: they only took persons who by talent or by training were especially capable of stating the phenomenon concerned through self-perception. In contrast to this Baldwin was interested in psychology as a means for diagnosis of psychic diseases and for testing human abilities. His strategy, therefore, was 'to test everybody' and 'to take persons just as they come'. The individual differences of reaction behaviour thus revealed, according to him, were not to be explained on the basis of a psychic apparatus but on the basis of individually different constitutions or acquired habits.

In this case, through the formulation of the experimental rules the different interests become the determinants of the object of scientific inquiry. Thus far, an example which may serve to illustrate what the very content of a constitution theory will be. Our investigations up to the present include the development of the general framework (more in detail for the part of quantification-processes) and some historical case studies. Constitution theory is to be conceived of as an historical epistemology. When it is performed it will present itself as a materialistic reconstruction of Kantian epistemology.

For the natural sciences today the question of how the object of a science is constituted is more of historical significance, but for the humanist sciences which so far still lack any foundation, this question is still a topical one. In the field of psychopathology, the various schools and paradigms differ in their view of what, in principle, is the object of their science and therapy. Thus behavioral modifications aimed at restituting the capacity for work postulate the isolated individual who is equipped with specific competencies or incompetencies to cope with his or her environment. Sociologically derived concepts on psychopathology, such as, for instance, some approaches in the research of schizophrenia, by contrast, thematize as the object of mental illness the forms of interaction the individual is involved in, for instance in the family. Psychoanalysis, in turn, oriented to the moral/amoral personality seeks to come to grips with the disturbed economy in the use of drives.

But even natural science as a whole must put up with being questioned by the constitution theory. For in the interest of technical manipulation, natural science has primarily sought out in nature those elementary building blocks and those laws of cause and effect which linearly connect intervention and success. Nature was the topic in as far as it could be put to use constructively. The interest in the existing order of nature and the insight into its meaning, which was the guiding principle for the Platonic natural sciences, has declined. At the present time, the consequences of the productive transformation of nature have made necessary the knowledge of the existing order of nature as well as the projection of new and more stable orders (ecology). In the long-term, this will give rise to a concept of nature which will enable humanity not only to understand itself as a product of nature but as an agent in nature.

4.2. Finalization

The thesis of finalization in science (31) is concerned with the substantive orientation of science to social problem situations and to social goals within mature sciences. The developmental theory of modern natural science of which the finalization thesis is a part, presupposes that natural science is, in principle, oriented to possible technological benefits but expects that particular disciplines will over longer periods have to ignore the latter, while putting emphasis on their own internal problem development, especially on the internal development of theory. In this phase then, according to Humboldt's thesis (32) the technical benefits will result only incidentally. But this process reaches completion, it is claimed; that is, the particular discipline reaches a state in which, in principle, it has a clear view of the field of phenomena, knows the elementary laws, and disposes of standardized experimental procedures. On the basis of this state of disciplinary maturity it then becomes possible to stimulate further development of the discipline, especially theoretical progress according to external problems.

The decisive point concerning the finalization thesis is that having a fundamental theory, that is a theory which 'in principle' describes a broad phenomenal field, by no means immediately makes it possible to work on the practical problems which are of interest in the particular field. Theory application is not a trivial business. Rather, it has turned out that to make a fundamental theory applicable to specific practical problems the former must be developed into a special theory oriented to these practical problems. In the terminology of Sneed's philosophy of science (33) this signifies: special theories are developed out of fundamental theories by means of kernel expansions. The concepts which engender these expansions are determined by intended applications which are of practical relevance. These concepts in many cases comprehend classes of boundary conditions, they are sometimes concepts of emergence which are necessary when research promotes to higher levels of complexity, they may be given by methods of approximation, which are determined by the particular conditions of the intended practical applications.

From the philosophy of science point of view, therefore, the finalization thesis is a theory on the application of theories. It shows that applications which have a practical interest through the production of special theories determine the content of a science. Putting it pragmatically, that is considering this from the point of view of its bearing on science planning, the finalization thesis signifies that within the development of a discipline at a specific point, namely the availability of a mature theory, it is justified to determine further development according to problems of application. But this requires criteria according to which it is possible to pass judgement on the maturity of a theory. We propose the following criteria to this end:

(1) it must be possible in a specific sense to state that the theory is 'complete'. (A theory would not be complete, for instance, if it had left undetermined the behaviour of particular variables to each other.)

(2) The theory must have had at least one successful application.

(3) One must have grounds for stating that the theory holds good with respect to fields to which it has not as yet been applied — in particular in the field of intended application (34).

Examples of finalization may be found wherever there exist well developed fundamental theories, i.e. since the nineteenth century to an increasing degree in all fields of physics and chemistry, and today, based on molecular biology, partly also in areas of biology and medical research. In terms of the internal goals of science, i.e. the search for fundamental laws, conceptual consistency, explanation of the phenomena, and the like, sub-disciplines here have come to a certain completeness. Their further theoretical development is determined by problems of application. To this end, concepts and procedures must be found which enable the fundamental theories to be actually applied, something that we have termed the 'fundaments for application' concepts. A characteristic example of this is the development of fluid mechanics after 1900; and a parallel can today be found in fusion research or in solid body research.

In the field of fluid mechanics there has existed, since about 1850, a mature theory: classical hydrodynamics. With it, this science had reached its internal goal, more than a hundred years ago, that was to find the fundamental laws for a determinate class of phenomena. But its development had not come to an end because this general theory was not applicable to most problems of practical interest (although there was no doubt about its validity in respect of these problems). What happened then was that a specific type of concepts was developed, notably by Ludwig Prandtl. These were no longer intended to provide conceptual fundaments for the phenomena the field entailed – this existed after all in the equations of classical hydro-dynamics – but rather the bases for *application*. The most famous of these is Prandtl's concept of the 'boundary layer' with which the general theory became applicable to problems of resistance in fluids of low viscosity. This meant that the practical problems arising in this field could be worked out. In the later history of fluid mechanics, there took place a differentiation into special theories which emerged as particular social interests arose: this is the case of the air-foil theory, the theory of the air-screw, the theory of lubrication, later aero-acoustics stimulated by the nuisance of aircraft noise, and today, it would seem in response to the spread of cardiac and circulatory diseases, there has risen a physiological theory of fluids (35).

Such examples as fluid mechanics (36), fusion research, and solid body research serve to highlight a very important phenomenon, which runs counter to the autonomist ideology and its sharp lines of separation between basic and applied research, that is that there can be basic research oriented to application. What counts here are not special applications, but 'application as such', that is the development of the conceptual tool-kit which makes it possible to apply a general theory to a practically determined class of problems.

The constitution theory and the finalization thesis are well suited to demostrate the connection of science and social goals at the origin of a discipline and, respectively, in the final phase of its development. They make evident that not every kind of knowledge is useable for purposes you like, that when one puts a stake only on one type of scientific knowledge it may appear that another kind of knowledge is lacking and that problems can be redefined, and particularly distorted, if they are thematized from a perspective of knowledge which originated in another problem-horizon. On the other hand, they do not make truth depend on interests. On the contrary, they demonstrate how the alternative norms assigned to scientific experience and discourse make possible a plurality in the search for truth. They make it evident that a social interest must exist not only relative to the utilization of knowledge but also relative to its production.

Our investigations in 'alternatives in science' took place mainly on a level of history and philosophy of science and of epistemology, that is of metascience. So they are no contributions for developing the alternatives needed themselves, but they may help to identify such alternatives, to destroy ideological restrictions and to open new prospects for science policy. Moreover, they may encourage a type of thinking which will relate science to human values in a more radical way. Let us now, in conclusion, turn to such ideas.

5. The Idea of the Good Science

The rise of modern science took place under the banner of the Baconian notion of the 'useful science'. Its opposite was not science that causes damage or worked harm but useless science, the kind of knowledge which only made sense as an occupation of the various schools of thought, or at best served to enlighten the individual. Within the framework of the Baconian conception, a science which served to perfect gunpowder was just as useful as a science which made possible the preservation of foodstuffs. The Baconian programme was realized on a scale which went far beyond any of the ideas of its author. But with that the antithesis of utility and harm have become separate, naiveté was dispelled, as if the kind of knowledge which served destruction could be in principle useful. Worse: even if this kind of knowledge was employed to do good, in the long term it could turn out to be harmful (37). This suggests that the ambivalence of modern science not only resides in the possibility of varying applications but in the structure of the knowledge it generates.

The question of the alternatives in science motivated by the critique of present-day science is inspired by the idea of a 'good science'. The idea of linking the ethical good with knowledge of the true is so far removed from today's philosophy of science that the claim that it is possible must draw for support on an historical realization of this idea, on Platonic science. With Plato virtue and knowledge, the true and the good are one. This was only possible because knowledge entailed knowing the order of the cosmos, virtue the orientation the individual found in this. Even Plato recognized that there was knowledge which was neutral in relation to the good, that is, 'technical' knowledge. That is the knowledge of the physician which can be used to cure or to kill, the knowledge of the builder, the military commander, the helmsman. But the practice of such knowledge is ambivalent even where according to the rules of the art it may be judged to be 'good', that is where it actually produces the work of art in conformity with the criteria accepted by the art.

This has been expressed by Plato most vividly in the dialogue 'Gorgias' when speaking about the pilot who has safely brought his ship-passengers to Athens, that is who has done what belonged to his competence:

> he who is the master of the art, and has done all this, gets out and walks about on the sea-shore by his ship an unassuming way. For he is able to reflect and is aware that he cannot tell which of his fellow-passengers he has benefitted, and which of them he has injured in not allowing them to be drowned. He knows that they are just the same when he has disembarked them as when they (512) embarked, and not a whit better either in their bodies or in their souls. (38)

Modern science, as a whole, seems to belong to this type of knowledge, that is to techne, the knowledge that is neutral in relation to the purposes for which it is used (39). Precisely this, as we have seen, the conservative ideologists of science acclaim as its virtue. The post-Platonic separation of good and true is acclaimed as the advance of rationality, the idea of good science discredited as regression.

If today attempts are made to realize the idea of the good science, then usually this is done by means of the social integration of science (socialist science), of social commitment (science for the people) or as the 'good scientist' (the responsibility of the scientist). The effort is guided by the image of the Hippocratic Oath to commit the application and even the production of knowledge to good (40), to exercise social control of the uses to which science is put , or to orient even the production of knowledge to social problems. As necessary as these attempts are, they remain exposed to objections which turn the intention underlying them into its opposite. These objections are based in the antagonistic or political character of the society: as long as rulers are arrayed against ruled, as long as states threaten other states, the idea of the good science cannot be realized by virtue of its social integration. The good science, from this point of view, presupposes the good society. And as long as the latter has not been realized, the idea of good science serves also to criticize society.

But if we take society as it is here and now and science as it is here and now, then the attempt to realize the idea of good science through the social integration of sciences faces a dilemma. If science is linked to social interests, it becomes particularistic in a way which seems to make it lose its scientific character (partisanship). If knowledge even here stays universal, then social integration is to no effect: science can then be used just as well for the opposite purpose. The objections show that the alternative consisting in social integration changes nothing unless science changes itself; that the idea of the good science must be realized at the level of cognition. Of course, we are, today, far away from being able to formulate this idea positively as the unity of the good and the true. Only the comparative realization of the idea of the good science becomes conceivable.

The experiences we have had with modern science, more particularly in this century, have shown that not every science or every knowledge can be used just as well for beneficial purposes as for harmful ones. Evidently, desturctive knowledge 'is easier to get'. The kinds of knowledge developed for warfare can not all equally and immediately be used for 'peaceful purposes'. On the contrary, there currently exists an immense disparity between the possibilities of destruction generated by scientific knowledge and the capacity of science to solve the problems of living with each other in peace. The dimension of the problems and the complexity of the knowledge required for this kind of problem-solving makes evident that the assertion about medical knowledge has always contained a delusion; let it pass that the knowledge which enables the physicians to cure people enables them to kill as well, the converse is not true: not every knowledge through which doctors can certainly kill people enables them to make someone healthy again or merely to keep them healthy. The idea of realizing the good science comparatively is based on the premise that the assertion about medical knowledge can, in the long run be disproved. The hope finds its support in studies of the epistemology and philosophy of science, such as constitution theory and the finalization in science thesis. They show that how an object is originally thematized, determines the way in which a science develops and that the intended practical applications are the determinants for the advancement of a mature science. If the goals we set ourselves determine the kind of science we have, then science will no longer be usable arbitrarily, for any kind of purpose.

Notes and References

1. J.R. Ravetz, 'Criticisms of Science'. In I. Spiegel-Rösing and D. de Solla-Price (Eds.), *Science, Technology and Society*, Sage, Beverley-Hills; 1977, pp. 71–89.
2. This critique could be based, e.g., on the hypothetical-deductive method itself: Science as well as mathematics cannot make sure its own fundaments. This is already Plato's argument in the State, Book 6, *cf* G. Böhme, 'Platons Theorie der exakten Wissenschaften', in *Antike and Abendland*, **XXII**, 1976, pp. 40–53.
3. M. Scheler, *Die Wissensformen und die Gesellschaft*, Bern, Müchen, Franke 2. Aufl. 1960
4. H. Marcuse, *Eros and Civilisation*, The Beacon Press, Boston, 1955.
5. The Anti-Science Movement had its main source within the academic youth 'retreating' from the scientific scene. It included a lot of heterogenous groups like hippies, green-culture groups, religious sects (christian but mainly oriental), yoga practitioners, psychodelics and other. It had its high tide in the late sixties and the early seventies. For some of the common objections to science see, for example, T. Roszak, *The Making of a Counter-Culture: Reflections on the Technocratic Society and its Youthful Opposition*, Faber and Faber, London, 1971; S. Cotgrove, 'Objections to Science', in *Nature* **250**, (1974) 764–767; S. Cotgrove, 'Anti-Science', in *New Scientist*, July 1973, p. 82.
6. J.R. Ravetz, *op cit.*, Note 1, 78.
7. G. Kröber, and H. Laitko, *Wissenschaft als soziale Kraft*, Akademic-Verlag, Berlin, 1976.
    > Our conviction that science in socialism has a greater potential in its impact on the development of the society and also as a socio-economic productive factor is based on such considerations. Within the context of the social relations of socialism any knowledge – even when it has been produced in the conditions of capitalism and thus has enriched the world fund of knowledge – has this potency and this objectively, irrespective of the intention of those who have produced such knowledge and the ideas of those who directly work with it.
8. The theory of the scientific and technical revolution shows close affinities with the ideas expressed by Marcuse in *Eros and Civilization*: For Marcuse, too, the development of natural science and technology leads to the satisfaction of the primal needs of man on the basis of a minimum of alienated labour, tendentially to the abolition of factory labour, to a change in human activity from productive to creative activity, to overcoming the realm of necessity and ultimately to achieving the realm of freedom.
9. In this section, we are concerned with contexts of reception which are restrictive against the idea of alternative science. But it should be mentioned that within Marxist thinking there exists a broad endeavour to turn from science criticism to 'new science', to socialist science. For a review of these endeavours see H. Rose and S. Rose, *The Radicalisation of Science*, Chapter 1, Macmillan, London, 1976.
10. H. Albert, 'Die Idee der Wahrheit und der Primat der Politik. Über die Konsequenzen der deutschen Ideologie für die Entwicklung der Wissenschaft'. In K. Hübner, N. Lobkowicz, H. Lübbe and G. Radnitzky (Eds.) *Die politische Herausforderung der Wissenschaft*, Hamburg: Hoffman und Campe, Hamburg, 1976, p. 15.
11. We find a similar motivation in Radnitzky's rejection of our thesis on the 'finaliza-

tion in science' formulated in 1976 (see G. Böhme, W.v.d. Daele, W. Krohn, *op. cit.*, Note 24). The thesis is perceived by him as an attempt to legitimize the control of science. Radnitzky writes *op. cit.*, Note 12: "The claim for political control of science is presented as an action which can be legitimized by science. Only in this way can the 'finalization theory' have any long-term effect on the public debate."

12. G. Radnitzky, 'Dogmatik und Skepsis: Folgen der Aufgabe der Wahrheitsidee für Wissenschaft und Politik'. In K. Hübner, N. Lobkowicz, H. Lübbe and G. Radnitzky (Eds.), *Die politische Herausforderung der Wissenschaft*, Hoffman und Campe Hamburg, 1976, p. 33
13. H. Poser, 'Wider die planende Destruktion'. In *Mitteilungen des Hochschulverbandes* **24**, 70 (1976).
14. Differentiation between truth and usefulness, mind and matter, man and nature, pure and applied science, science and technology.
15. *Cf.* the writings by Carl Schmitt, but also by Scheler *op. cit.,* Note 3 (e.g. 32) and H. Schelsky, *Die Arbeit tun die anderen*, Westdeutscher Verlag, Opladen, 2. Aufl. 1975.
16. This additional connection is clearly perceived by the conservative ideologists of science: Radnitzky writes: "Ultimately one comes to recognize that the 'finalization theory' is nothing else but one of the many politically inspired fairy tales intended to justify ideological claims to power, which are spread around by parts of our reasoning intelligentia (analyzed by Schelzky 1975)", *op. cit.*, Note 12, 31.
17. J.R. Ravetz, *op. cit.*, Note 1, 22a: "Popper's genuine scientist does not induce from phenomena or verify hypotheses; for that way lies pseudo-science and the self-deceptions of the astrologer, the Freudian and the Marxist".
18. E. Mendelsohn, 'The Social Construction of Scientific Knowledge'. In E. Mendelsohn, P. Weingart and R. Whitley (Eds.), *The Social Production of Scientific Knowledge*, D. Reidel, Dordrecht, Boston, 1974, pp. 3–26.
19. W. van den Daele, 'The Social Construction of Science: Institutionalization and Definition of Positive Science in the Latter Half of the Seventeenth Century'. In E. Mendelsohn, P. Weingart, and R. Whitley (Eds.), *The Social Production of Scientific Knowledge*, D. Reidel, Dordrecht, Boston, 1977, pp. 27–54.
20. I can't give here a short concept of what science in China as an alternative of modern science is or was, but see the works of J. Needham and his collaborators. The problem I find with his studies is, that he does not deal with Chinese science on its own merits but always follows the question 'what is China's contribution to universal science?' As a consequence in Germany a collection of Needham's articles appeared under the title 'Scientific Universalism'. In his article on 'History and Human Values: A Chinese Perspective for World Science and Technology' where Needham explicitly puts the question what may be expected from China regarding the recent forms of science criticism he concludes: "Science is a unity, and it is not different in China from anywhere else, but what we can rightly object to is the idea that science is the only valid way of apprehending the universe", in H. Rose and S. Rose, *The Radicalisation of Science*, Macmillan, London, 1976, pp. 90–117.
21. G. Böhme, 'Platons Theorie der exakten Wissenschaften'. In *Antike und Abendland* **XXII**, 40–53 (1976).

22. G. Böhme, 'Ist Goethes Farbenlehre Wissenschaft?' In *Studia Leibnitiana* **IX**, 27–54 (1977).
22a. G. Böhme, '1848 und die Nicht-Enstehung der Sozialmedizin'. In *Kennis en Methode*, 1979.
23. T. S. Kuhn, *The Structure of Scientific Revolutions*. 2nd edn. enlarged. In O. Neurath and R. Carnap (Eds.), *Foundations of the Unity of Science*, Vol. II. Chicago London, University Press, 1970.
24. G. Böhme, W.v.d. Daele, W. Krohn, 'Alternatives in Science', in *Int. J. Sociology* **VIII**, 70–94 (1978); 'Finalization in Science', in *Social Science Information* **XV**, 307–330. (1976).
25. Contemporary protophysics has clearly dissociated itself from this standpoint.
26. It must be recalled to mind that this is the case also in the Kantian constitution theory, Kant's radical formulations notwithstanding (the understanding is itself the law-giver of nature, the source of the laws of nature, Kant's *Critique of Pure Reason* A127). Qualities such as warmth, weight, hardness and the like and primary forces must be *given* to the understanding through the senses.
27. G. Böhme, 'Quantifizierung – Metrisierung. Versuch einer Unterscheidung erkenntnistheoretischer und wissenschaftstheoretischer Moments in Prozeß der Bildung von quantitativen Begriffen'. In *Zeitschrift für allgemeine Wissenschaftstheorie*, **VII**, 209–222 (1976).

G. Böhme 'Quantifizierung und Instrumentenentwicklung. Zur Beziehung der Entwicklung wissenschaftlicher Begriffsbildung und Meßtechnik'. In *Technikgeschichte* **43**, 307–313 (1976).
28. *Cf.* Toulmin's study of the various concepts of 'natural motion' and the correlated concepts of the phenomena requiring explanation, in S. Toulmin, *Foresight and Understanding. An enquiry into the aims of science*, Hutchinson, London, 1961.
29. G. Böhme, 'Cognitive Norms, Knowledge Interests and the Constitution of the Scientific Object: A Case Study in the Functioning of Rules for Experimentation'. In E. Mendelsohn, P. Weingart and R. Whitley (Eds.), *The Social Production of Scientific Knowledge*, D. Reidel, Dordrecht, Boston, 1977, 129–141.
30. G. Böhme, 'Die Ausdifferenzierung wissenschaftlicher Diskurse'. In N. Stehr and R. König, *Wissenschaftssoziologie, Sonderheft 18 der KZfSS*, Köln, 1975, pp. 231–253.
31. G. Böhme, W. van den Daele and W. Krohn, 'Finalization in Science'. In *Social Science Information* **XV**, 307–330 (1976).
32. "Science often is richest in its blessings for life when it seems to withdraw from life".
33. J.B. Sneed, *The Logical Structure of Mathematical Physics*, D. Reidel, Dordrecht, 1971.
34. G. Böhme, W. v.d. Daele and R. Hohlfeld, 'Finalisierung revisited'. In G. Böhme, W. v.d. Daele, R. Hohlfeld, W. Krohn and T. Spengler, *Die gesellschaftliche Orientierung wissenschaftlichen Fortschritts*, Suhrkamp, Frankfurt, 1978.
35. G. Böhme, 'Autonomisierung und Finalisierung. Ein Vergleich der Theorienentwicklung in der Gärungsforschung und der Strömungsforschung. In *op. cit.*, Note 34.
36. Further case studies related to the finalization thesis are to be found in *op. cit.*, Note 34. For a critical evaluation of the thesis in English see R. Johnston, 'Finaliza-

tion. A new start for science policy?', in *Social Science Information* **XV**, 31–336 (1976); J.M.D. Symes, 'Policy and Maturity in Science', in *Social Science Information* **XV**, 337–347 (1976).
37. This refers to such instances as the demographic explosion as a consequence of the progress of modern medicine and the use of DDT to combat pests and insects.
38. 'Gorgias' 511e, 512a, transl. by B. Jowett.
39. The defining feature of modern science since its origin is precisely the linkage with the crafts. On the other hand, the studies which follow up the rise of modern science (Scheler *op. cit.* Note 3, E. Zilsel, *Die sozialen Ursprünge der neuzeitlichen Wissenschaft*, hrsg. und übers. von W. Krohn, Suhrkamp Frankfurt, 1976) make evident that the influence of speculative philosophy was just as important. For it was the speculative philosophy which left to modern science an interest in pure theory which represents a critical potential in contrast to the merely instrumental utilization of science.
40. Thus, for several years a group of molecular geneticists has been working out, and sought to get adopted, a code of rules for the handling of genes in the test-tube experiments. These rules concern not only work with genes but genetic engineering as such. In the United States of America and in Great Britain these rules have in the meanwhile become binding for publicly-funded research organizations. In the Federal Republic of Germany a similar arrangement will soon be put into effect.

COUNTER-MOVEMENTS AND THE SCIENCES: THESES SUPPORTING COUNTER-MOVEMENTS TO THE 'SCIENTISATION OF THE WORLD'

OTTO ULLRICH
Freie Universität, Berlin

1. Counter-movements to the Industrial System

Oppositional movements against the 'rationality' of industrial society, or at least an uneasy feeling in the face of industrial civilization have been with the 'modern age' right from the beginning. These counter-movements were interpreted by the mainstream of development as the rearguard actions of 'reactionary' minorities; as a 'critique of civilization' by a few philosophers who deplored the loss of privilege by a declining class, or as a 'storming of the machines' by a few craftsmen who lacked proper historical foresight. There was general conviction that, as history progressed, science and technology would progressively unfold the productive forces sweeping away any doubts, because the rationality of the new production methods would bring about undreamed-of advantages for everyone, only provided that they spread widely and rapidly to embrace all areas of production and reproduction.

Max Weber argued that the sciences, industrialization and bureaucratization were carrying rationalization into every aspect of our lives. Not only has rationalization become much more widespread and intensive since then, but also the counter-movements and the uneasy feeling about industrial civilization's rationality have assumed new forms and proportions — particularly over the last decade. We cannot look any more upon these counter-movements merely as rearguard actions or as backward-looking nostalgia. Quite possibly they foreshadow new ways of life in a post-industrial society.

At present it is still difficult to discern one leading and dominant line in the heterogeneous manifestations of these counter-movements in Europe, the U.S. and Japan, in the manifold manifestations of protest: from citizens' initiatives against government planning over the heads of those concerned, through the alternative movements in rural and urban communes, the womens' movement, to the counter-culture of adolescents; it is still difficult to assess how these counter-movements could prevail *politically* against the ruling powers and structures of the capitalist industrial system.

There are, however, some indications that the counter-movements could gather force. Three of these indications are: damage caused by the industrial system can be directly experienced; many young people have no perspectives, the system finds itself in a crisis of legitimacy.

The likelihood that people experiencing detrimental effects from the industrial system will develop some consciousness of their causes is increasing. Massive environmental pollution, for instance, has been around since industrialization, but mainly in the environment of the lower strata of the population. And they had to aim, in their political struggle, for economic improvement, and thus for the expansion of the industrial system. Today the effects of industrialization encroach increasingly upon the total life environment of the middle classes. This, along with greater economic independence and better education, is the reason why today most counter-movements to industrial civilization are carried by middle class people. The threat the industrial system poses to the human environment becomes increasingly more total and local confinement of its effects is more and more difficult, whilst the proportion of what were once middle class characteristics within the population, such as education and relative economic independence, is increasing. Hence, counter-movements to industrial civilization might well gain a broader basis, even if for the time being such protest will in many cases only be an ecologically oriented 'ersatz critique' and not, as yet, a fundamental critique of the capitalist industrial system.

The industrial system itself produces a further, increasing potential for protest: young people faced with dull work which they cannot control, lacking training in the professions or crafts or having work inadequate to their training, faced with unemployment and regimentation through the authorities, find that the system offers them less and less of a satisfactory life-prospect. Why should not this increasing potential (after all the aberrations and attempts to escape into the thrills of speed and into drugs) also become a creative potential for alternative ways of life, particularly since a large proportion of young people — for instance due to enforced unemployment — has been somewhat less crippled than the older generations by the work ethic of the industrial civilization?

Another fact that might strongly support this development is the crisis of legitimacy of the entire industrial system. The industrial system has been legitimated in a secular way by the fact that, for generations, those in power were able to create a noticeable rise in the incomes of the population. They achieved this by distributing merely part of the *increase* in production and at the same moment succeeded in distracting attention from the system's unchanging wrongs and inequalities. Therefore, the 'ruling elements', together with the unions, are doggedly trying to coax the system into further growth. But for many reasons growth — at least in Western industrial nations — has its limits. Thus, in the struggle over distribution the focus can shift to power and riches, and in this dispute the rationality of the entire industrial system might be challenged.

Until now, social scientists have analysed some of the counter-movements to industrial civilization from some distance. They noted that these movements have a tendency towards mysticism, towards idyllic set-ups, that they are apolitical, and so forth. But the sociologists themselves mostly kept out of the processes involved. It is in this connection I would like to present the first of my theses.

THESIS 1. If it is true that due to structural factors and developmental trends in overdeveloped industrial systems counter-movements to industrial civilization will not merely remain marginal phenomena, then there is a need for social scientists to actively participate in these counter-movements *as* social scientists.

This thesis is based on the following considerations: to ensure continued meaningful survival of humanity, post-modern, post-industrial societies will have to develop. Why should sociologists confine themselves to describing this process after the event, speculating about it or even hampering it by providing expert opinions for those in power? During certain phases, counter-movements might be in need of the knowledge and the reflective capacities of sociologists if they are to continue to develop meaningfully.

However, before sociologists can appropriately decide how to transmit their knowledge within counter-movements, they themselves will first have to go through an educational process. Within the counter-movements they will have to train their practical imagination for alternative ways of life and, on a theoretical plane, they will have to free themselves more widely and more thoroughly of the metaphysics on which the industrial system is based. A key issue which may reflect such metaphysics, which is central to the rationality of the industrial system, and which has direct bearing on the prevailing social sciences, is the rationality of the sciences. Therefore, I present my second thesis.

THESIS 2. Before social scientists can sensibly participate as social scientists in counter-movements against industrial civilization, they will have to break radically with the prevailing concept of the rationality of science.

2. The Metaphysics which underlies the Idea that the 'Scientisation' of the World is Possible

Nearly all activities that claim to be sciences follow the pattern of the natural sciences. Since the emergence of the natural sciences about 300 years ago, scientists have, with mounting success, endeavoured to interpret (natural) processes in accordance with (mathematical) principles. This is achieved only by isolating a specific process, taking it out of its natural context, and reconstructing it within an experimental setting in such a way that the desired process — predicted according to a given principle — takes place in a controlled and reproducible way. This elementary experimental situation, then, serves as the pattern on which industry is subsequently built up step by step: objects and processes that are encountered are analysed as far as this is possible, that is, they are broken down into their parts, and then synthesized according to some extraneous principle and re-assembled for the purposes of an external purpose. According to this concept of the rationality of the sciences, the 'scientisation of the world' means that the world is taken apart and put together again in such a way that all its angles are 'straightened', that all its spontaneity, self-willedness and fortuity are eliminated, that all processes are predictable and can be planned, monitored, and controlled centrally.

This rationality, which later became the rationality of the industrial system, is based on a metaphysics which is specific, a product of history, and has to be recognized as such. The emergence of the modern sciences has to be seen against the background of the long, preparatory and persistent tradition of the Judeo-Christian interpretation of the world (1). Within the Judeo-Christian tradition an anthropocentric attitude specific to this culture evolved. It saw humanity indissolubly bound to God by covenant, while the rest of the world was seen as something separate, serving only as the background against which humanity had to prove itself to fulfill divine revelation. The Judeo-Christian search for God's scheme of salvation was continued in the sciences when theological methods ceased to yield satisfactory results. The early scientists, with hardly any exception, saw it as their task to study in nature and in its law God's plan and intention (2). The divine revelation was continued by humanity (via the scientist), hoping thus to gain insight into the divine purposes of the natural order of the world. Although this hope was to be disappointed, and for most natural scientists God was no longer an explicit issue, the assumption of nature's rationality which lay at the very basis of this search for God has persisted to this day (3).

We shall not list here the numerous effects of this concept of rationality on all areas of modern life and thought. It will still take us some time to free ourselves from the compulsions of this rationality in our thinking and in the reality created by men (4). Nevertheless, since it is a key element for the subject in hand, I would like to highlight one aspect of this rationality: the metaphysical assumption that nature can be fully investigated and is rational, is also the metaphysical basis of all planning concepts aiming at the total permeation of society.

When it became evident that the scheme of salvation could not be deciphered from nature, the belief was only strengthened that humanity could create *itself* – provided one kept to the scientific method used up to then to enquire into God's Reason in nature. Thus the divine scheme was not only to be known to humanity, to the scientist, but it was to be completed by humanity. And this scheme could not be ambivalent, could not leave open any alternative for the potential whims, fancies and interests of some people; it had to be unequivocal, the only correct and reasonable one, as befits a divine plan. This ideology of the 'only best way' is at the basis of all 'scientific' social models and of all technocratic planning concepts. The metaphysicians of planning are convinced that, if the various elements and their operating mechanisms are only researched sufficiently thoroughly after the scientific methods, they can be assembled and organized in such a way that the whole will become the Reasonable which, strictly speaking, will have to be accepted by all. Whoever amongst those concerned does not share this view is either not yet scientifically enlightened enough, malevolent or deranged. Thus even Saint-Simon believed that if the scientific method is extensively used, political conflicts of interest, for instance, are unnecessary and only objects would have to be administered, in line with scientific considerations.

In spite of the great number of critical voices which have meanwhile been raised against this metaphysics of rationality, of a divine scheme that can be realized, it will be a long time yet before this calamitous metaphysical concept will be broken down. Within "real socialism" this metaphysics has been condensed into the concept of 'scientific socialism', to form a hermetic *weltanschauung*. In that camp, the certainty of eventual salvation is intact. Although the capitalist camp produces more diverse ideologies, the predominant expectation of salvation, of a final settlement of all crises rests on the total 'scientisation' of the world (5). And, in both camps, scientific progress, the unfolding of the productive forces towards ever greater perfection, is both the driving force of the actual process of making the world ever more scientific, and the hope of salvation from all evil, to government, capital, parties, unions, and as yet still to the majority of the population. The theses discussed in the following section present a number of reasons why this hope is becoming increasingly fallacious.

3. Reasons for Increasing Urgency of a Counter-movement to the 'Scientisation' of the World

THESIS 3. There is increasing divergence between productive capability through scientific technology and the ability to assume control and responsibility.

It is appropriate to the rational character of the sciences to reconstruct the world in isolated portions according to ideal principles. This implies that the scientist's interest is focussed on such an isolated process. A scientist's interest and horizon are necessarily 'limited' and as science progresses this limitation increases, since increasing division of labour allows a steadily decreasing part of the whole to be viewed by any one scientist. If there was a 'scheme of salvation' at the base of the whole undertaking, we could, in spite of any individual's limited view, rest assured that the findings and scientific products — be it new substances, equipment, technologies — would, in the end, fit together to form a whole benevolent towards humans. But we cannot build on this scheme of salvation, and this means that the sum total of the rational partial activities of limited scientists can become an irrational whole which threatens humanity in its entirety.

This threat has become real since the sciences have generated large-scale equipment and technologies. The scientists who develop new substances, equipment and processes are not interested in the effects these produce beyond the horizon of their scientific interests, how they are used, and whether they are useful to humanity. In many cases, it would not be possible to investigate these effects, anyway, because we have no knowledge of the manifold effects and cannot obtain it, since they radiate far into space and time.

The ecological effects of, for instance, artificially synthesized chemicals which do not 'fit' into the system whose balance has been achieved through evolution over many millions of years can hardly be fully determined in advance, since it is impossible to reproduce the earth in the laboratory. Necessarily, we make do with partial experiments in the laboratory and then experiment upon the earth itself. Thus, it is quite usual to determine afterwards, and often too late, how dangerous the effects of a new substance, a new process, are for the ecological system. Examination of the effects is further hampered by possible time-lag, as many of them show up only after a very long time and, only appear as the cumulative effect of many small influences which then cannot be traced back individually. How, for instance, packaging, chemical food additives or, air pollution — even when below the legal limits — will affect the genes of humans and other organisms will probably never be 'scientifically' and fully disclosed. Nevertheless, such effects *could* be catastrophic.

The divergence between scientifically created facts and their consequences is even more obvious in nuclear physics and nuclear technology. Even the effects to the system of additional low levels of radiation, which are passed off as tolerable, have not been scientifically and exhaustively 'researched', nor, in principle, can they be. In addition, nuclear technology has consequences affecting the future of many hundreds of generations. If, moreover, any sizable accident occurs — which is not impossible, after all — this very future could be destroyed. Nobody *can* really take responsibility for these realities (6).

Here, too, the continuing certainty of having entered an indissoluble covenant with God, the unbounded optimism that salvation will result through new inventions solving all problems is possibly what induces many scientists to continue without concern for the direction taken by the sciences and technology, in the belief that they are completing the Creation (7).

The belief that it is possible for man to complete and improve on Creation is particularly pronounced in one of the more recent sciences: genetics. Scientists here seriously believe that the manipulation of genes is feasible and can be controlled by the human brain. This would in effect mean that a very central intervention in the reproductive mechanism of a system which it took many millions of years to stabilize could be controlled by an organism which in itself is only one of the products — possibly not even the best outcome — of this evolution. As long as this anthropocentric megalomania existed only in the spiritual form of religion or philosophy it may have been bearable. Today it commands the technological potential of actual material intervention and is already producing irreversible changes in the ecological system. Today, therefore, there is a need to struggle against all offshoots of this religion. Any further 'scientisation' can only accelerate this perilous process. A counter-movement to the "scientisation" of the world which is based on the expectation of salvation, is vitally *necessary* for the survival of the system we need to live in.

THESIS 4. Sciences are increasingly focussing on areas remote from life and threatening life.

For many reasons, science concentrates increasingly on large-scale technological projects. Seen from the point of view of the scientists, large-scale technological projects are an ideal field for testing and demonstrating a model of a world totally given over to science, the scientist's and engineer's world. Here nothing must be left to chance, everything must be planned and ordered, following the logic of a consistent and unequivocal rationality. Realization of such a project involves the solution of many problems which are particularly numerous and difficult in large-scale projects. It is just this, however, that challenges scientists and engineers alike and therefore their interest is focussed on such projects to an extraordinary degree.

On the other hand, large-scale technological projects to an ever increasing extent form the basis on which new scientific findings are produced. In nearly all spheres of science larger and larger apparatus, ever more encompassing forms of organizing activities involving division of labour, are needed to produce still further 'progress' on the road taken.

Such large-scale technological projects in which science tends to concentrate and reproduce itself, are typically indifferent or hostile towards the vital interests of the population. This is quite obvious in the case of the large-scale armaments projects which all over the world employ the great majority of scientists in extremely interesting work. But it also applies to most other large-scale projects which are passed off as cornerstones of progress, such as space projects, supersonic aircraft, nuclear power stations, particle accelerators, etc.. The benefits of these projects — if any — affect only a few privileged people. Basically they serve the concept of national prestige harboured by governments and the profit interests of corporations. They place an exceptional strain on the environment, and scientific knowledge, the deeper insights into the 'natural order of the world' gained from them only serve to give satisfaction to the scientists themselves. Any communication of this 'knowledge' to the paying population has long been discontinued. On the contrary, scientists ask that their activities be met with faith equal to that accorded to shamans and priests.

It is no coincidence that the most concrete forms of counter-movements to the industrial system are kindled by such large-scale technological projects. Very often, the members of citizens' initiative groups can gain immediate experience of the threat to life which these progressive projects pose, but, to date, their protest has not been very effective in the face of the rationality and power these systems are creating.

THESIS 5. Science and technology increasingly serve to maintain existing power.

As discussed above, humans engaged in the sciences in modern times have approached nature from a power motive. According to the biblical mandate to 'subdue the earth', the anthropocentric scientist sees nature only as an object to be controlled and exploited. The specific method of reconstructing nature via an experimental situation has led towards perfect control. The rationality of a linear-hierarchical-causal linkage, which originally was assumed to be the pattern of the divine structure of the world, has now become the knitting pattern for all projects, since humanity has taken Creation into its own hands. Large-scale technological projects, production systems, institutions, and bureaucracies, in line with scientific rationality, are set up to allow smooth control from outside. All disturbing influences which could, for instance, arise from the spontaneity, self-will or independence of humans have been 'straightened out' or 'smoothed out', everything is ordered according to a rational plan. The humans 'built into' such a system are condemned to subordinacy and to suppress spontaneous interests of their own. On this basis, the exploitative interests of ruling classes and groupings can be presented through science and technology as 'factual laws' which must be followed inevitably.

For this and many other reasons, the reality created through scientific rationality in industrial civilizations is eminently suitable to maintain, justify and obscure an 'objectivized' rule (8). Also, industrial technology serves to secure for the industrial states new dependencies within a new technological colonialism.

If follows that if we want individuals and peoples to decide their paths and their actions themselves and in line with democratic principles, there will have to be much less 'scientific order' in this world.

THESIS 6. The social, psychic and physical cost of a 'scientific world' is becoming less and less bearable.

The other side of the anthropocentric attitude is that humans, while controlling and exploiting nature, while briskly unfolding scientific and technological progress, have repressed the fact that they themselves are part of nature and that, by exploiting and transforming nature, they are mutilating themselves. As a product of evolution, humans, too, are optimally adapted to certain environments. Despite the breadth over which these environments can vary, some limits remain which cannot be transcended for any length of time without creating damage. This is obvious in the case of nutrition. But also, for instance, a certain intensity and accumulation of stressors such as noise or permanent repression of one's own impulses to act create psychic and physical disorders. The industrial system as a whole is developing increasingly into an environment unfavourable to the human species, incompatible with humans, making them psychically and physically ill. Take just the rhythm of the industrial system: the great speed of the processes, the high and the fast rates of turn-over of utensils and information, of knowledge acquired and of professional activities, the shifting and elimination of the day/night cycle, and so forth. Quite apart from all other factors, the industrial system's rhythm, continuously accelerated by scientific progress (amongst others) is highly incompatible with a sensible 'rhythm of life', which more and more people want.

The industrial system does violence not only to our external, but also to our internal, nature and it is no coincidence that all those different and scattered counter-movements to industrial civilization have one common denominator: they are all searching for a new relationship with nature.

But we do not need to venture into the concept of 'human nature' — so tricky for sociologists — to highlight the industrial system's inadequacy for humans. It is sufficient to correctly interpret the observable symptoms of diseases.

In production, for instance, the large degree of division of labour, the suppression of individual interests, the prefabricated, straightened work paths, the dependence on set conditions outside the individual's control, all add up to create a motivational crisis and, in many cases, a disposition to psychic disorders. Together with the influences from the other deformed spheres of life, the industrial system creates an alarming number of psychically ill people, and attempts to heal these by scientific psychiatry, encounter-movements or group-dynamical activities are but hopeless endeavours.

The total social cost in the great achievements of our industrial age has, as yet, not been computed and certainly will be difficult to assess, since comparative criteria are often lacking. How do you, for instance, 'book' the fact that today the high degree of division of labour has made people more dependent on each other — is this a welcome 'socialization' of human activities or a lamentable loss of the individual's capacity to be autonomous, independent?

In many cases it would be sufficient to use existing criteria to recognize the irrationality and the great social cost of industrial civilization's achievements. Slowly, everybody must come to understand that most social patterns and facilities needed for a society cannot sensibly be established along the lines of a factory system: with regional concentration, high degree of division of labour, standardization of individual activities and organizational concentration through central management. This supposedly effective factory system has been introduced into all areas: production and distribution of commodities and victuals, education and training of children and adolescents, health care and the care and 'keeping' of the old, housing and recreation.

It will be necessary to study extensively these areas one by one so as to be able to compare the material, social and psychic gains and costs. In the area of health care, for instance, there is increasing evidence and findings are accumulating that the highly sophisticated, scientific medical system produces more sick persons than it cures or protects from disease, and that the picture is the more unfavourable the more science and scientific technology are built into the system. Also, in areas which normally are not regarded as part of the industrial system, such as agriculture, closer inspection will show that here, too, the use of typically fragmented rational sciences and technologies has done more harm than good. Counting everything: the cost of fertilizers, the destruction of the soil's regenerative capacities, the cost of pesticides, the free-of-charge natural pesticides thus eliminated (e.g. birds), the large amounts of energy used in modern agricultural production, the consequential medical cost of chemically treated food, the work time and environmental cost of the chemical, machine-building and energy supply industries, etc., a comparison of the total productivity of agriculture using scientific technology with that of a less 'scientific' mode of production shows that using scientific technology the *overall* balance is less favourable (9).

4. On the Self-understanding of a Social Science Wishing to Criticize the Trend towards the 'Scientisation' of the World

A social science critical of the sciences must make every effort to free itself from the spell of the very rationality it criticizes. This demands that it ceases to strive to imitate the natural sciences and takes a critical view of all the usual characteristics of a 'scientific approach', such as faculties defined by a high degree of division of labour, by development of a terminology specific to the field and understood with difficulty even by fellow-experts, and by a strict methodology. Social scientists who wish to break away from the currently predominant scientific activity and to take part in counter-movements as social scientists must above all be able to impart the results of the analyses in intelligible language to those concerned. Social science will be relevant to the counter-movements to industrial civilization when it is capable of communication with the group members of the counter-movements. For this process of communication it will also he important to note that it is impossible to overcome the industrial system merely by negating existing facts and circumstances, merely by criticizing instrumental reason. Alternative institutions and ways of life must be tried out. It seems obvious that sociologists, too, — even if many of them will first need to go through an educational process within the counter-movements — can help with such alternative institutional suggestions and that in the long run they will probably have to do so, since social scientists — at least according to their claim — have acquired the knowledge and ability to reflect on psychic and social mechanisms. And, together with historical knowledge, this can prevent counter-movements from repeating certain mistakes and from re-entering culs-de-sac. For this it will neither be necessary to develop self-contained and consistent scientific theories, nor will it be sensible to subject the work to any strict method (10). In any event, teaching and preaching as outsiders with a know-all attitude, so typical of dogmatic strategies of liberation through a scheme of ultimate salvation, will have to be strictly avoided.

In order to avoid any misunderstandings, it would seem sensible not to call such a social science 'science' any more. But there is no other word for a human undertaking which orders within the framework of language experiences and desires that can be examined and shared, so that this order is instructive, prepares action and is open to criticism of new experiences and new desires and thus is differentiated and marked off against all closed 'language games' of religious conceptions of the world. Therefore, I shall retain the term 'science'. For the game of language that is science, the minimum conditions worth retaining are: the possibility to communicate thoughts and theorems by means of 'reasonable' discourses between equal individuals (inter-individual communicability) and to verify or refute these statements through such discourses. In this sense, social science critical of the sciences must remain a 'science' and must not become a new doctrine of salvation. Moreover, by retaining the term science we highlight the fact that criticism of the industrial system cannot be meaningful if it rejects *all* scientific results, methods and processes. Only a radical *review* of all scientific results, methods and processes is meaningful. Thus, by way of summary, my seventh thesis follows.

THESIS 7. Any social science critical of the industrial system must simultaneously meet several preconditions if it is to be relevant to the social changes needed: It must study extensively and criticize existing social conditions in the industrial system, it must be able to submit concrete institutional alternative proposals, it must put the critique, the analysis and the proposals into a language that can be communicated to those concerned, and it must remain science in the sense of an open language game that can be criticized and deals with statements that can be verified inter-individually.

5. Central Themes for a Social Science Wishing to Support Counter-movements to the Industrial System

The theses put forward in Section 3 suggest not only the need for counter-movements to a further scientisation of the world but also the themes of any social science wishing to participate in these counter-movements. Now, I would like to touch upon four field which address typical characteristics of the industrial system, which are linked by a problem running right through the industrial society, and which lend themselves to social scientific treatment.

The problem running through industrial society is the problem of the individual's participation in its processes and activities. In line with rationality of the industrial system, people, in the last analysis, are disturbing factors within the scientifically conceived processes. Their individual specific idiosyncrasies, interests, desires to participate, as well as satisfactory work situations must, according to the prevailing tendency, be removed by rationalizing them. In this system, the majority is only being used — it is not the designer and driving force of these processes. There are four characteristics of the industrial society which in their specific form systematically hamper individuals from participating in their own destinies. These are: division of labour, magnitude of enterprises, degree of centralization and necessity of growth. All four characteristics are enhanced by the dominant interests within the industrial system, and, in a very specific manner, also by the rationality of scientific/technological progress.

Ivan Illich's thesis that many human institutions and inventions have socially critical limits, where initial advantages turn into disadvantages, would seem to hold true. For the four 'inventions' just mentioned, these socially critical limits most certainly do exist. They will have to be located in each case by exact study and analysis and experiments. There are numerous indications that in the overdeveloped industrial societies these limits have for sometime clearly been exceeded in the four fields just mentioned, and that it is, in any event, now a question of contraction. By way of conclusion I shall — again in the form of annotated theses — deal in somewhat greater detail with these four areas.

THESIS 8. The limits of sensible *division of labour* have been exceeded in the industrial societies.

Division of labour is an important cultural invention enabling humans to achieve more in cooperation than isolated individuals could do. But division of labour, excessive as it is today, only serves the interests of a few. For the many it has more disadvantages than advantages: it fragments the work, which then cannot give any satisfaction, it thus creates unnecessary psychic costs and makes it more difficult for the individual to have a view of the whole; in many cases, it makes it impossible for individuals to have any influence on the direction of the processes in which they are involved, etc. It will be essential to backtrack on division of labour to the point where individuals in self-determined communities of production controllable by themselves are in a position to design and produce complete utensils and use them. At least part of the utensils or of the food must be lifted out of the process of production of commodities with its highly advanced division of labour, and intensive reliance on the sciences and on capital. This would, to some degree, give back another dimension to people's qualifications. Qualifications would not only be a marketable commodity, to be used only within an anonymous labour market in other people's interests, indifferent towards the specific substance of the work on hand; they would be qualifications and work which, at least within the individual's sector, would enhance his or her feeling of own value and opportunities for self-realization; and it would enable people — while still sharing and dividing labour — to live as a cooperative community of autonomous individuals.

THESIS 9. When organizations of production or technological projects exceed a certain *size*, they become dysfunctional for the people concerned. In the industrial nations institutions and facilities in many cases exceed their functional size and this is reason enough to encourage without reservations any struggle against large scale projects.

As the size of a technological project increases, so does the necessity for the division of labour, of more hierarchical organization and of control; the uncontrolled tendency of projects/institutions to assume a life of their own increases, as does the prospect of uncalled-for effects. Accordingly, the likelihood of satisfactory work in these projects and of a democratic participation in the control diminishes on 'objective' grounds. There are many reasons why, in an industrial society, projects become increasingly larger (11). The rationality of science, too, aims at larger projects, so that the interests which are a driving force in the intensification of the scientific approach are also the interests that hinder democratization and the opportunity for satisfactory work for the majority.

That exporting large-scale technological plants to so-called undeveloped countries is an obstacle to any meaningful chances of development for these countries has been recognized (12). For the overdeveloped countries, critical sociology of organization should find and clearly define the socially critical limit to large-scale projects. Thus, for instance, the measure of control necessitated by a certain magnitude of organization should be considered as a social cost factor. Certain dangerous scientific developments and technologies should become intolerable for a society as soon as the mere control requirements become too great, as for instance, through surveillance systems, police protection, protection against potential sabotage, limitation of civil rights, and so on.

The question of the magnitude of projects and organizations also touches upon another important issue that has not been solved by the industrial system: how to achieve solidarity between people in the industrial society? Freedom and equality could possibly be created within an industrial system by 'organizational' means; at least it is possible to create an accepted illusion that freedom and equality can be achieved in the industrial system. The third demand of the bourgeois revolution, fraternity, is linked more closely to human experience and emotion. Therefore it is less easily replaced by illusions or organized from above. But above all it contradicts in a more fundamental way the rationality of the industrial system. In large organizations directing all efforts towards a certain aim, for instance, the nature of the system will not leave any room for modes of behaviour that might carry interhuman solidarity. Typically, attempts to realize the ideal of solidarity within the industrial system have, therefore, remained limited to counter-movements. One of the prerequisites for increasing the chances of such behaviour in a post-industrial society will be a reduction in size of most of the institutions of industrial civilization.

THESIS 10. To begin with, any *centralization* must be suspected of increasing the opportunities of dominance for a few and of increasing the dependence of many. Therefore, all movements for autonomy and all efforts towards decentralization deserve encourgement from a sociological point of view.

In modern times, the article of faith that joining, uniting mergers, consolidations and centralization are to be regarded as progress is closely linked with the metaphysics which holds that human societies can be organized according to a uniform principle of reason. It is not only the processes that actually take place in economics and politics that have been directed towards ever increasing concentration and centralization, but also most social Utopias for socialist and communist societies, in the modern age, imply that increasing centralization is progress. From a certain level of centralization onward, however, the only and exclusive winners in this process are parasitic central powers.

The sciences today regard the central powers — whether they be combines or governments — as their natural allies. For reasons indicated in Thesis 4 above, the natural sciences require an increasingly high capital input to achieve further 'progress'. The central powers control the surplus of production and will use it to promote the sciences and technology, if only for reasons of self-preservation (see Thesis 5). The interests that have organized themselves in the established scientific activities and that are propelling the sciences and technology along this path, quite rightly feel that financing their projects by the central powers is the most convenient way of evading the need for legitimation by the paying population.

If science is to contribute towards a richer life and not, as is mostly the case today, towards the jeopardization and destruction of our lives, it must become possible for the population to agree to concrete scientific projects.

One precondition for this is that the sciences be disconnected from the central powers and that existing decision-making structures be radically decentralized.

THESIS 11. Further growth of production in the industrial nations serves mainly purely to maintain the drudgery of meaningless work and irreversibly endangers the ecological system.

We have already mentioned to what extent growth is the vital core of the industrial system. Growth is closely linked with the syndrome of rationality in our modern world, with the anthropocentric approach of the restless search for salvation: the earth is the object to be worked upon, the field within which Humanity and humans have to prove themselves, where by constant efforts and labour the individual can prove before God and humans that he/she counts among the chosen. The sciences were part of this effort and today are amongst the essential driving forces of this restless striving for ever greater growth, for ever new products and processes, for ever more complete exploitation of the earth's resources. The sciences are so much interlaced with the idea of growth that science without this characteristic is difficult to conceive of, and a scientifically oriented world will probably always be a world of growth fetishism.

In this connection, and in conclusion, I would like to add a remark on the evaluation of counter-movements. Formerly the most important counter-movement in the capitalist industrial system has been tied to this system of restless labour for more and new things: the labour movement as organized in unions. Coupled via the controlling mechanism to receive a share merely of the increase in production, each existing and new job, regardless of whether it is for armaments, for technology that is an environmental hazard, or for socially meaningless products, becomes a source of income, a source that is worth preserving. In this way, the unionized industrial labour force is turned into a mainstay of the industrial system.

The most difficult and most important task ahead, in which sociologists, too, will have to help, will thus be to suggest to the unions and the industrial labour force a realizable and acceptable alternative programme which will lead them out of the cul-de-sac of production/growth/job fetishism.

This will be difficult, because it touches upon the entire body of industrial reproductive mechanisms and relations of dominance, while, at the same time, those concerned are not as yet aware of any necessity to change their way of life in principle, since they have too deeply internalized the philosophies, attitudes, behaviour, etc., considered 'natural' in industrial civilization.

This work will be important because counter-movements for alternative ways of life beyond the industrial system will only be accepted by society if the organized industrial labour force participates in them.

Notes and References

1. See, e.g., Hans Blumenberg, *Die Genesis der kopernikanischen Welt*, Frankfurt/Main, 1975.
2. See, e.g., F.E. Manuel, *The Religion of Isaac Newton*, London, 1974; or the religio-sociological works of Max Weber.
3. To give just one *explicit* example of many similar formulations by contemporary physicians:
    > It is the basic creed of science that the physical world has been created on a consistent and integral plan. We therefore expect that all our observations, however unconnected they may appear to us, must be part of this pattern. The discovery of this pattern is the sole object of all fundamental research.

    Thus the eminent low temperature scientist Professor Kurt Mendelssohn in a 1966 textbook. (K. Mendelssohn: *The Quest for Absolute Zero. The Meaning of Low Temperature Physics*, World University Library, Weidenfeld and Nicolson, London, 1966, p. 14.)
4. For attempts to achieve this freedom of thought see, e.g., the works of Magoroh Maruyama on post-industrial logic.
5. *Cf.*, for instance, the high hopes Daniel Bell placed in the 'scientisation' of a post-industrial world: Daniel Bell, *The Coming of Post-Industrial Society*, New York, 1973.
6. Very early, and emphatically, Guenther Anders pointed out the danger of the discrepancy between ability to create and ability to take responsibility: G. Anders, *Die Antiquiertheit des Menschen. Über die Seele im Zeitalter der zweiten industriellen Revolution*, München, 1956.
7. To avoid any misunderstanding, we have to insert here a qualifying note. The deciding and driving factors of the scientific process are by no means limited to the metaphysical motive, to this certainty of salvation. Scientific projects, such as the nuclear technology mentioned, arose from complex structural interactions of the sciences, state and capital, which cannot with any claim to accuracy be described within the scope of a paper (I tried to analyse these in (8)). I have highlighted in this paper the metaphysical background of the rationality of industrial civilization, because in most critical analyses of capital this aspect is not seen and thus, too often, a critique of capitalism fails to become critique of the industrial system.
8. This connection is described in greater detail in Otto Ullrich, *Technik und Herrschaft, Vom Handwerk zur verdinglichten Blockstruktur industrieller Produktion*,

Frankfurt/Main, 1977. There, the other Theses of this Section, too, are dealt with in detail. Moreover, they are placed in a wider context there, which, for instance also takes into consideration the influence of capital.
9. See, e.g., Barry Commoner, *The Closing Circle*, New York, 1971.
10. See Paul K. Feyerabend, *Against Method*, New Left Books, London, 1975.
11. See Thesis 4 and more detailed also in (8).
12. See, e.g., E.F. Schumacher, *Small is Beautiful*, London, 1973.

SCIENCE AND IGNORANCE

MICHAEL GRUPP

Synopsis, Institute de Recherche Alternative, Lodève (Hérault), France

> I know that I am ignorant
> Socrates

Prologue

Imagine the funeral of science. The relatives (the scientific community) follow the casket of science, crying over lost funding and unemployment. The undertaker (the Government) wrings his hands in false compassion, while counting the savings that will relieve the hard-pressed budget. The vultures: ever-growing gangs of quacks, thriving on pseudoscientific vocabulary. Finally, the heirs: ideologies of all kinds that fill the immense space left by science in the hearts of men — the need to believe.

However, although there is some anticipatory action in the funeral parlour, and the procession is lining up, practising for the great day, science is not dead. On the contrary, like many ailing old men it is still at the peak of its power.

The Power of Science

It is a commonplace statement that never before has science had a greater impact on the life of all. Technology is but one aspect of this, for scientific reasoning is invoked in support of all kinds of decisions concerning everyone.

On the other hand, science is not everybody's game. It is specialized and coded up to a point where scientists often do not understand their own work. In addition, science is played behind closed doors, ranging from degrees required for scientific jobs to armed guards around nuclear research facilities.

Distrust in science is regarded as obscurantism and is even against the law, as in the case of compulsory medical treatment. Therefore most people trust in science, this thing they know nothing about. They are wrong.

In the case of technology, science is called on to make predictions for two different kinds of questions. The first kind is, 'How can we do this?', 'this' being any limited problem such as the production of electricity, disease control, etc. The ability of science to cope with this kind of problem was the reason for its success. I will call this kind of problem 'simple'.

The other kind of question is, 'what will happen if we do this on a large scale?', the question of so-called 'side-effects'. I will call this kind of problem 'complex', and will try to show that the ability of science to cope with it is very limited (1).

Simple Problems and the Scientific Experiment

Simple Problems are limited to a small number of parameters. Let us take an example. When President Kennedy asked his advisors how supersonic transport could be achieved, the answer dealt with payload, speed, fuel consumption, range, noise, prices, each of these parameters being dependent on all the others (2). It can be seen that this answer reduced the number of possible answer in two ways: first, it considered only technical aspects; second, it retained of all technical aspects only the ones that seemed most important at the time. This example shows an important feature of Simple Problems. For simplicity's sake, they eclipse a part of reality; they have this feature in common with the scientific experiment:

Here I am, the subject, wanting to find out something about the object — say, an atom. This is not easy, because my atom is not alone. There are many atoms, of different elements, moving at different speeds; there are magnetic and electric fields, all kinds of radiation — and there I am, too. As C. F. von Weizsäcker puts it, "the subject ... lives itself in the world of the objects as one of its parts" (3). There goes the neat distinction between subject and object. Thus, if I don't want to change my topic (the atom) and start to enquire about the universe, including myself, I have to start eclipsing parts of reality (4). This process can be divided into two steps:

(1) First, I will eclipse myself (the 'cut'). In order to reconstitute an object, I draw a line (the 'cut') between the 'myself' and the 'world around me'. The

location of this cut is arbitrary; it can be located between my measuring apparatus and the atom, or at the limits of my body, or between my eyes and my brain, or even further back. The theory of measurement (5) postulates that the location of the cut (within certain limits) does not influence the result of the experiment.

This concept has an obvious advantage. Before the cut, there is probably only one scientific answer to the riddles of nature: 'Since I don't know anything about myself, how can I know anything about anything?', and even this answer contains, in the words 'I' and 'myself', a sort of cut. The cut creates an artificial ('objective') reality that differs from 'real' reality in one respect: 'I' am not part of it. Science then claims this objective reality as its field. It leaves the realm on 'my' side, the subjective part of the world, to philosophy – a dubious heritage. All explanatory systems seem to run into paradoxes, from Kant's a priori judgements (where do they come from?) over Hegel's phenomenologic self-reflection (the never-ending circle) to the rude 'I-don't-care-science-is-O.K.' of older positivism (is science O.K.?) (6).

The cut also has serious disadvantages. By definition, it excepts a part of reality, the subject, from scientific reflection. This part of reality can be made as small as the progress of science allows, but it will always be there. This situation is a modern version of Plato's cave. We are looking at nature through an instrument with unknown characteristics – embarrassing for a scientist.

Finally, the cut leaves me with a strange view of the world. The universe on one side, 'me' on the other. Or 'me' as the centre of the universe? Wildly speculating, does science by its own logic perpetuate the archaic anthropocentric and egocentric views that made its existence possible? And what would this mean for its results, for its intentions, and for its effects?

(2) After having myself eclipsed from the object space theoretically, I will start to restrict this space practically. Remember, there are still the other atoms bumping into 'my' atom, there are the fields and radiations, there is a room, a house, etc., and all of these things have an influence on my atom, however weak. Since I don't want to do research on this whole zoo of different physical phenomena at the same time (I would not even find my atom in the middle of all this mess), I will have to cut out some of these outside influences (e.g. by putting my atom into a 'vacuum'), and I will have to keep some of these inflences under control. Ideally, this should apply to all

outside influences. This process is called the preparation of a well-defined initial state. After having prepared my initial state, I can start playing around with some of the outside influences, the parameters, to study their effect on my atom. One of the golden rules-of-thumb of experimentation states that it is unwise to vary more than one parameter at the time to avoid confusion. As a consequence, the number of parameters studied in an experiment must be small.

Like the cut, the preparation of an initial state is a necessary condition for any scientific experiment. Clearly, it can never be complete; the good experimenter is the one who keeps his eye on *all important* (as opposed to *all*, which would be impossible) outside influences.

Seen from this angle, the scientific experiment seems to be a pretty shaky method, and nobody seems to really understand how and why it works, but it does work. Certainly, the conclusions drawn from any particular experiment have a good chance of being wrong. Of the few scientists (including myself) whose work I know in some detail, I don't know one who doesn't have some skeletons in the closet, publications containing more or less serious errors. I know some scientists who have mostly skeletons in their closet, and who would certainly regard the citation of most of their earlier publications as impertinent. Since scientists are well-educated people, very few of these errors go public, even within the scientific community. However, the conclusions of different scientists concerning the same problem tend to converge into a consensus. This consensus might not be definite, it can always be stirred up by new findings; but many of these consensus have proved quite stable, and have permitted far-reaching and (so far) correct predictions. This evident success of science, together with philosophy's as evident inability to explain it, have caused the positivist school to disregard philosophy, and to postulate instead an inherent tendency of science toward truth (7).

We have seen that the scientific experiment is a method adapted to problems with a small number of parameters. This can explain why science deals very successfully with Simple Problems (8).

However, science does not have a monopoly on Simple Problems. Most of the known solutions to these problems were (and probably still are) found by simple trial and error over the generations. Science is certainly the quicker way towards a solution, avoiding most of the errors necessary with trial and error. On the other hand, just by being quicker, science leaves less time for feedback. Any doubts on the how and why of a possible solution are shorted out by the very swiftness of science's proceeding. A scientist's solution to a Simple Problem might be very different from a solution arrived at by trial and error (remember, though there is convergence, the 'final' solution might take a long time to reach). In fact, contemporary science seems to have an inherent tendency to propose centralized, highly complex solutions (9).

Complex Problems

Let us return to the example of supersonic transport, and let us assume that the Simple Problem has been solved. We now have a prototype and we know how many people can fly how fast for what price. Then somebody comes up and asks: 'What will happen once 500 of these planes fly around?'.

This is a new type of question: the Complex Problem. The Simple Problem was concerned with payload, speed, fuel consumption etc. ('etc.' being a very few additional parameters). Nothing else was of interest then. Now the question is unlimited. Possible answers to it could concern the effects of fluorocarbons and other combustion products on the stratosphere, the effects of the sonic boom on life and structures on the surface, or the changes this kind of transport will cause to the organization of societies using it or not using it (10).

There are many more issues addressed by this Complex Problem. I could try to enumerate all of them, but nobody can guarantee that I have not forgotten the most important ones. This is the first difficulty encountered by a scientist who faces a Complex Problem. No method, however perfect, can yield answers to answers to unknown questions.

Further, some of the issues involved might appear more complex than the Complex Problem we started off with. Take, for example, the case of possible change in the structure of society caused by supersonic transport. For whatever reason, problems of this kind have so far eluded all rigorous treatment.

We are little better off with issues like the effects on the stratosphere. Certainly, this problem can be subdivided into a great number of Simple Problems, most of them already solved, but the knowledge of laws of thermodynamics, of intermolecular potentials, etc., together with some empirical rules, is not enough to describe such a big and complicated mixture of gases.

A comparison of today's weather with yesterday's forecast usually gives a good example for this.

Complex Problems do not always *look* complex. Moreover, many Simple Problems cease to be simple once the solution has to be very precise. Take the example of the safety of a wooden bridge. Let us assume that you have built such a bridge and, before crossing it for the first time, you want to know how safe it is. You could tackle this problem by trial and error, but you might have difficulties finding someone who is willing to try. So you turn to science, read a book on statics, figure out how much weight your bridge can take under the given circumstances and, if this weight is sensibly bigger than yours, you step on to your bridge — hoping that the book, your computation, and the materials used are free of hidden flaws: a typical example of a Simple Problem.

Let us then assume that you made it over the bridge all right. Now, you want to open the bridge for the public, and again you want to know how safe it is. This time, the answer has to be much more precise. Before, while you were on your own, you might have accepted a chance of one in a thousand to go down with the bridge, at least for the first try. So you concentrated your study on the most evident (and perhaps the most important) accident cause, the collapse of the bridge caused by simple overload. Now, with millions of people who want to cross the bridge over many years, the situation has changed. First of all, a risk of one in a thousand would now mean thousands of accidents, and would therefore be unacceptable. Therefore, you now have to analyze even quite unlikely events. Further, the use of the bridge over the years by many people creates other causes of risk. There is wear, rot, fire, weather, possible earthquakes and floods, but also panic, vibrations caused by people marching in step, drunks who don't realize that they are on a bridge, and many other possible accident causes, including combinations of several single causes. Now you are faced with a Complex Problem.

Luckily for you, your don't have to solve it ab initio. All you have to do is to look up the accident statistics of wooden bridges. If you find these statistics published (which is rather unlikely for objects of some industrial interest), they will give you a good idea of the number and type of accidents that will be caused by your bridge. Reality, by the way of statistics of past events, has discerned the important contributors to the risk from the unimportant. If these important contributors to risk are Simple Problems themselves, the Complex Problem can be reduced to a number of Simple Problems.

If past experience is not there to 'guide' the solution, Complex Problems will remain complex. Parts of them might still be broken up into Simple Problems, and these Simple Problems might be solved, but there is no way of telling what has been forgotten in the process. The problem of safety of nuclear power plants gives an example: in principle it is the same problem as the wooden bridge, though somewhat more complicated. The difference is that there is no sufficient background of experience to guide the solution. As a consequence, even the most elaborate safety studies leave a doubt that lies beyond error or dishonesty: they are not complete, they can never be complete. What is missing? (11).

In spite of a great number of investigated details, science can only yield correct solutions of Complex Problems by looking back. Science is not able to encompass such problems exhaustively, which is what would be needed for a reasonable forecast. However, after reality has shown the outcome of a Complex Problem, science is in many cases quite capable of discerning the important causes that have led to this outcome. R. Pirsig calls this science's "20/20 hindsight" (12). This hindsight can be of use the next time the same (or a reasonably similar) problem comes around.

Science as Religion

The conclusion is simple. Any scientist who firmly predicts the outcome of a Complex Problem that has not had a well-studied and reasonably similar parallel in the past is a quack. If his prediction is used by decision-makers to justify their decisions, science takes the classical role of religion: to back the rulers' decisions with higher (and always obscure) knowledge.

Science has an important advantage over religion. Changes in a religion's dogma usually take a long time to be accepted by believers, whereas science calls the detection of past errors its progress. This makes science more versatile than religion.

Science has also its promise of paradise: tomorrow's better world of perfect technology. It is instructive to read predictions made by popular technical magazines in the 1950s for the 1970s. I remember an illustrated article about future housing. In the 'typical modern house of the 1970s' described, most of the household work is done by machines while the family relaxes in easy chairs. An open door to the outside shows lush green, a friendly jungle. As for the Simple Problems, the machines, the prediction is fairly good, though technical progress has been somewhat slower than predicted. As for the Complex Problems, the prediction is naïve. First of all, nobody is at home to relax during the day (unemployment excepted . . .): the husband's work day has not become shorter, as implicitly predicted. The wife is at work, too. 'Her' dishwasher saves work washing dishes, but it has to be built, repaired, powered, and replaced. The wife now 'washes her dishes' at work. Also, looking out of typical modern housing, we don't see lush green, but a city from high up.

The prospect of technical progress can be very attractive, but this progress often creates more and more severe problems than it solves. For a while, technical progress can be re-proposed to solve the new problems, too. It is like the waiter with a tray full of glasses. He drops one, catches it in mid-air with a sudden movement that cuases two more glasses to fall. His next decision will be decisive for the rest of the tray. For this reason, the attractiveness of science as a religion is wearing off rapidly.

There are two main factors that have made the misuse of science as a religion possible: the scientists themselves, and the organization of science.

The Education of a Scientist

During the first years of his studies, a student of science mainly learns one thing: to conceal his incomprehension. This incomprehension is not due to a lack of intelligence (the student doesn't know this). It is mainly due to the new 'language' of science; but, even once this new language is understood, the facts expressed cannot be understood in depth because they cannot be connected to other known facts. It is important for a beginner student not to show his incomprehension in order to appear as a good student. This can go to extremes: I remember my first course in theoretical physics. It was a pretty bad course, and my comprehension of it was about nil. So was, as far as I can tell, the other students'. At one moment a courageous student got up and said that the course was going too fast for him. He asked the professor to slow down and to explain in more detail. The professor, in turn, asked the audience to vote on that. There were only a handful of students who admitted that they did not follow. The great majority of the students claimed that they agreed with the speed of the course, and some students even said that the course was too slow (I was one of them). What we learned in this way was not physics, but bluff. Students who show their incomprehension are usually among the first drop-outs, leaving room for the others who are more inclined to bluff and less sensitive for their own ignorance: the believers.

This selection is negative. The best dam-builder is not the one who knows the most details about dam-building, but the one who knows best what he doesn't know. Take the latest dam failure in the United States. Instead of thorough probing, assumptions were made about the permeability of the ground the dam was built on. These assumptions proved to be false and the dam, perfectly built otherwise, was carried away by the water during the first test.

Apart from this education to bluff, scientific curricula have several other shortcomings.

First, science is not presented as a method, but as a system of facts. "University teaching . . . is not a preparation for research: science is taught in a dogmatic way, like a truth revealed" (13). (The style of textbooks illustrates this. In textbooks, reality always 'obeys the law of . . .', which is obviously wrong. In the best of all possible cases, the inverse is true and the law obeys reality.) During my whole time at university, there was no systematic course or book on how to perform an experiment, how to check if a source of information is trustworthy, or on the necessary distrust in scientific equipment. This way, the student learns an unfounded trust in scientific results and develops a warped picture of scientific methodology.

Second, there is no information about the influence of bias and personal interest on scientific results. Lacking this information, scientific results can be influenced even without the researchers being aware. All one has to do is to preserve some scientists from unemployment and to pay them well. They will almost certainly judge their own work as interesting, safe and necessary. Moreover, most of them will believe in their own objectivity. They don't know better.

Third, the student learns to restrict his scope. Again textbooks: examples start with 'given . . .' and, even if what follows is obvious nonsense, the students are not invited to apply their intelligence on it. They learn to blank out parts of reality and to start thinking exactly what they are told.

Last, and most important, science is not presented within its (extrascientific) contexts. At least two of these, the philosophical and the cultural context, are essential for the understanding of science. In spite of this, it is possible to study any of the natural sciences today without the knowledge that there are other ways than science to look at nature, that there are cultures, different from ours, explaining nature by theories different from ours (14).

The same is true for the philosophical context. What makes knowledge possible? What is the relation between the world outside and its impressions on our senses? How well are the objects in this world outside fitted by our (naïve or scientific) categories? Do these categories bias our image of the world? Disregard for questions of this type leads to blind belief of science in itself, to a view of humanity that starts with Galileo and ends at the borders of the developed countries.

In many important respects, scientific curricula keep the students in the dark. The students learn how to look through a microscope instead of looking around themselves. If this education succeeds, the scientists produced by it will be virtually blind. They will need guidance, and they will get it.

The Organization of Science

Science by itself does not produce goods. A scientist has to be appreciated by others in order to be fed; he or she is dependent on funding. This way, science is controlled even in the rather rare cases when the scientist does not work under explicit orders concerning the object of his research. In spite of the freedom of science being guaranteed in most constitutions and fundamental laws, this control is rather tight. In times of changing priorities for funding, a scientist can do either of three things: resist (continue the work and perish slowly), obey (change the object), or cheat (continue the work and call it by another name). Cheating is not very difficult; science can be obscure, and obscura can be called by any name. However, most scientists choose to obey, sometimes after a period of cheating; scientists who are hooked onto their object by lasting interest are rare. No wonder. Laypeople will never be able to understand in depth how boring scientific work can be. Progress on the 'borders of knowledge' is usually very slow, there are few experiments performed today that actually decide about the value of a theory: the famous experimentum crucis is part of a past history, with few exceptions. So, it is usually hypocrisy that makes scientists present their field in the form: 'I am interested in . . . '. For many scientists, doing science is a game, fun to play. The object itself does not matter, so why not give in to the pressure to change it?

Further, most scientific work is done by highly specialized groups with very little communication between them. As we have seen, scientists are trained not to cross the borders of their specialty, except when told. Organized this way, science can be used to fill in details and numbers into any scheme decided in advance. Such a scheme can be, and usually is, decided on non-scientific grounds; it will still be sold under the label of science.

The Danger

This way, science has been invoked lately to justify many important decisions and trends. Many people identify science with increasingly complex technology, growing specialization of production and centralization of decision.

No scientist can take the responsibility for these trends on scientific grounds: we have seen above that science cannot make reasonable forecasts about the overall effects of Complex Problems. Thus, humanity is not guided into the future by its leaders of knowledge, but led into the night. Possible outcomes range from a more comfortable, reasonable and peaceful world to a world organized like a galley ship, rowing straight into the ultimate ecological disaster.

There is much at stake. If science keeps taking sides without sufficient, or against better, knowledge, led by the petty interest of its own evergrowing importance and by the fear to lose privileges, it might make way for a new obscurantism. A major disaster caused by one of science's pets could shatter all confidence in science and could lead to a dark hate of reason.

Ignorance

There is one obvious way to avoid such a development — that is, to play science by the rules. Of course there is no set of rules for science acknowledged by all scientists, but on one rule (among others) there would be unanimous consent — at least in theory: 'Where the pre-scientific thinker is unable to confess ignorance on any question of vital practical import, the good scientist is always ready to do so' (14). In short: don't bluff.

Speaking more practically, it is necessary to show that not all scientists agree on the solutions presented in the name of science. It is easy to defend such a disagreement in a scientific debate, since the 'official' positions have to pretend complete (or sufficient) knowledge of Complex Problems. The nuclear debate has shown this; the 'officials' frequently felt forced to retaliate by non-scientific means ranging from disciplinary measures against opponents to redefinitions of science as a game only for insiders (15).

On the other hand, positive counterposition — alternative solutions — are far more difficult to formulate. Much research has to be done, research that cannot be limited by borders of scientific disciplines or by the interest of the funding agencies. In other words, such research can hardly be done within existing research institutions. New interdisciplinary and explicitly independent possibilities for research have to be created.

Further, alternative solutions might not be accepted easily; they will not have the false assurance of crash programmes. No universal solutions, cheap and quick, will be offered. More likely, alternative solutions will only make careful and conditional forecasts and will be based on trial and error. They will start on a small scale, analyze the outcome, be corrected, and continue carefully. If sound, these solutions will reflect true science: a humble method to find out about nature.

Before it is too late, the use of science should be taken out of the hands of those who have made a religion out of it. Science's High Priests are false.

Epilogue

Once upon a time, there was a young physicist. He worked in a nuclear research facility, studying completely esoteric problems, writing papers that were — like all his colleagues' papers — never read. Day by day he grew more tired of this artificial life, of the silent collaboration of his colleagues (including himself) who justified by their very existence the follies committed in their name. He came to the conclusion that this collaboration was not only due to a blinding education and to the 'system', but also due to personal interest: there was much unemployment, and science made a well-paid job. His own view, he deemed just; was it not contrary to these interests and, thus, independent? So he decided to quit his career. He went out to the country with his friends (16) to build an independent research institute. One winter day, while writing an article on the necessity of independent research, he remembered a French proverb saying "everybody preaches for his own church". He suddenly realized that he was no exception.

Notes and References

1. The distinction between Simple and Complex Problems is by no means precise. It is meant to be a descriptive distinction, like big and small.
2. For the history of the debate over supersonic transport, see J. Primack and F.v. Hippel, *Advice and Dissent*, New York, 1974.
3. C.F. von Weizsäcker, *Die Einheit der Natur*, München, 1971, p. 140 (translation of the cited passage by the author).
4. It might appear strange to dissect concepts like subject and object on one hand, and to stick with the concept of reality on the other; but it seems to me that further discussion on reality – though essential for serious philosophy as opposed to this article – will not get us any further in this context. Let's do what – mostly – everybody does, and assume that 'it is there'.
5. See, e.g., A. Messiah, *Quantum Mechanics*, Amsterdam, 1965.
6. See, e.g., J. Habermas, *Erkenntnis und Interesse*, Frankfurt, 1973.
7. See, e.g., C.S. Peirce in *Collected Papers*, Hartshorne and Seiss (Eds.), VIII, 12, citation in ref. 6.
8. There is another common point between science and Simple Problems, which is the convergence of solutions. E.F. Schumacher (*Quest*, Vol 1, No. 4, 1977) has pointed this out on the example of the bicycle: in spite of a considerable effort in research and development, the bicycle has stayed the same for the last 60 years, minor improvements excepted. Of course, parallel to the case of conclusions from a scientific experiment, there is no way of proving that there will be no fundamental improvements in the future.
9. O. Ullrich (*Technik und Herrschaft*, Frankfurt, 1977) explains this tendency with the industry's general interest in the development of such a technology. To my experience, this is not the whole story. Scientists are fabulous clients for dealers in scientific equipment. They don't spend their own money, and part of their prestige is proportional to the sum of money they spend for their work. Thus, you can sell most anything to a scientist, provided it is complicated, expensive, and quickly outdated. This fact has certainly an influence on the solutions proposed by science.
10. See I. Illich, *Energy and Equity*, London, 1974.
11. In fact, reality is catching up. The most serious accident so far in a nuclear power plant was due to fire. Fire had not been identified before as a dangerous accident initiating event. See M. Grupp in, Bericht der Diskussionsgruppe 5, *Informations-Kampagne Kernenergie*, (Eds.) Bundespressedienst, Wien, 1977.
12. R. Pirsig, *Zen and the Art of Motorcycle Maintenance*, New York, 1974.
13. La Redaction de "Survivre", in *(Auto) critique de le Science*, J.M. Lévy-Leblond

and A. Jaubert (Eds.), Paris, 1975 (translation of the cited passage by the author).
14. See, e.g., R. Horton, *African Traditional Thought and Western Science*, Africa, Vol. XXXVII, 1967. Horton explains that the structure of traditional African thought shows many parallels to the structure of modern theoretical thinking.
15. E. Münch and P. Borsch, *Atomwirtschaft*, 27, 1977.
16. This paper is dedicated to them. While I write this, comfortably seated in an unheated room (but inside), they work outside in the cold, building our Institute.

MANFRED E. A. SCHMUTZER

Interdisziplinäres Forschungszentrum: Technik, Naturwissenschaft, Gesellschaft, Wien

> ... even though we may offer the Giver of Life,
> emeralds and fine ointments,
> if with the offering of necklaces you are invoked,
> with the strength of the eagle, of the tiger,
> **IT MAY BE THAT ON EARTH NO ONE SPEAKS THE TRUTH.**
>
> Cantares Mexicanos (fol. 25/V)
Nahuatl Philosophy

> Oh, come with old Khayyám, and leave the Wise
> To talk; one thing is certain, that life flies;
> One thing is certain and the Rest is Lies;
> The Flower that once has blown for ever dies.
>
> The Rubáiyat of Omar Khayyám
First Edition, verse XXVI

> Useless to reply that reality, too, is ordered.
> It may be so, but in accordance with divine laws —
> I translate: inhuman laws — which we will never
> completely perceive. Tlön may be a labyrinth,
> but it is a labyrinth plotted by men, a labyrinth
> destined to be deciphered by men.
>
> Tlön, Uqbar, Orbis Tertius
Jorge Luis Borges

'... It May Be That On Earth No-one Speaks the Truth'

Hundreds of thousands of library shelves bend under the load of books written on topics of the philosophy of science, sociology of science, history of science and their like, boasting answers to questions of the type of 'What is science?', 'How did it emerge from the dawn of history?', 'How does it work?', etc. It seems an idle enterprise to add to this legion of well-conceived, critical and uncritical accounts a further weight of unread paper.

What apology may I offer for wasting paper and printer's ink, and the energy and time of both the reader (if he or she should actually exist) and myself. Being neither Sir, nor Professor, nor even a member of any of the illustrious circles of academics and scientists, I can put little forward in my defence, other than to point to the pleasure that I find in writing and in reading my own words in print, particularly if these words are soaked with the deep seated distrust I feel when looking from a distance and from outside at the products and performance of the scientific community.

The distrust of the outsider leads me to play alternative little games, like jigsaw puzzles, with little sticks of scientific knowledge. My own little games create my own little rules and lead me in time to hide some little sticks – shame on me – or to ask for some more little sticks, which in time again I cannot find on the bending shelves of the libraries. Woe to me, shame and scandal to the scientific family.

But missing links and hidden sticks put generously aside – they are to be taken for granted here and there, in my own games and in yours – what is it that makes me so suspicious about the great, official games? I believe it is simply that your games are not my games. Both of us play, however, more or less with the same sticks. It must not be thought at this point that I am possibly afraid of losing some of my sticks to you. Our sticks are, as we all know well, of a material that can be divided many times without growing smaller. One might try to express this in scientific language by writing:

$$a/n = a \text{ for all } n = 1, 2, 3, \ldots$$
$$0 < a < \infty$$

Now I suppose that even quite untrained sociologists will look at least a little puzzled, when glancing at the formula and some might even raise an eyebrow in doubt or protest. Is my statement wrong, or is it your maths? Shall I invite you to go to look it up on the shelves? Admittedly, it is a triviality, not worth quarrelling with. Your maths is right, just as much as my statement is; what I should have done is abstain from applying your maths to my statement. This solution resolves our potential conflict, but to me it expresses another riddle. How do we know what we should apply when?

I presume at this point some of my unlikely readers might reply; yeah, this is precisely what science is all about: the scientist is taught what to apply, and what not through empirical experience. Now, calculating the trajectory of a ballistic missile or the output of a combustion engine or of a steam turbine cannot be done by simple application of thermodynamics or the law of gravity (1). Yet despite the well-known and necessary adjustments of thermodynamics or of the law of gravity, when they are applied, these theories are still hailed as embodiments of scientific knowledge. Again it is not easy to know what to apply and when.

This brings me back to my previous remark about my own little games. What is the difference between them and the great, serious games of scientists? Can we define science as an institution producing knowledge according to an agreed methodology? Are my little games not scientific because they lack social backing?

A serious objection to this definition is found in the assertions of scholars like Feyerabend, who claims that no unique methodology of the sciences exists (2). Feyerabend's claim is provocative and controversial. But even when retreating to less controversial views it proves difficult to define the specificities of the sciences. Erwin Schrödinger (3), for instance, understands the following two principal assumptions as constituting the sciences. One is that nature is comprehensible. The other is that nature is objective, i.e. it is independent of the observing subject (4). Both principles, as Schrödinger himself stresses, are conventions with definite consequences. One consequence of the first principle is the need for an additional concept to causality, namely, random or chance, which has to take care of those phenomenological experiences not lending themselves to causal explanations.

A consequence of the second convention is even more perplexing. It is intrinsically connected with the scientific method. Excluding the observer consequently means excluding his or her sense organs, which necessitates the introduction of impersonal senses, measurement instruments, reacting objectively to objective inputs. Accepting those devices forms a minimal social convention. Beyond this convention, however, their acceptance does something to our joint experience; it atomizes experience, as any measurement is discrete. Borrowing from Georgescu-Roegen's terminology, one might say science creates an arithmomorphic continuum (5), with an inexhaustible number of gaps in between little droplets of agreed reality. Contrary to subjective experience, scientific experience is scattered. The tension between subjective and objective experience produces demands for interpolation, for filling the gaps. How many buildings can be made from a heap of bricks? Many, quite different ones. This does not mean all at once, as the construction of one might inhibit the construction of another; but the number of possible constructions is, however, large (6).

From these considerations, we are free to deduce that science is a method of producing and destroying knowledge and world images on the basis of some social conventions. Change one or more of these conventions and you will change knowledge and even facts. The construction of world images is, as even Schrödinger argues, part of all sciences, not just of historical disciplines. The resulting world images are, due to their atomistic factuality, in principle, negotiable (7). The fact that a number of different houses may be constructed out of a large number of bricks, leads us as sociologists to look for a sociological perspective, which allows to comprehend why preference is given to one kind of world image rather than to another.

There is a thread which leads through history, at least from the days of Friar Occam, via Nicholas Copernicus to Kirchhoff, Mach and Popper, who formulated principles of an economy of scientific thought, sometimes also called parsimony. It says that out of two or more possible explanations the least demanding should be chosen. How does economy come into nature? Why should it be the least demanding explanation? According to our daily experience it is hard to accept that nature's vast range of virtues includes parsimonious thinking, less so as even the human economy is far from being economical or parsimonious (8). Neither in the biosphere nor in the universe does nature exhibit deep concern for economy. Arguing the other way round, declaring present day economic principles as manifestations of natural principles is not more convincing as the two do not match. It is we who attribute to nature qualities which we value highly at present. It is the fetishism of scarcity which governs the thoughts of our ruling strata (to avoid another term) which is projected into nature and brought to our minds as a natural phenomenon through science.

It is a miracle, but we only tend to find what we look for. It is therefore a legitimate question to ask, what will our world image be like, when scarcity ceases to be the fetish of economists, or even better when economists cease to be dominant in our societies?

The principle of parsimony indicates which additional ingredients give the real flavour to the scientific stew. In science we are confronted with an ancient tradition, the tradition of *projecting social principles into the heavens*. This thought is hardly novel, but related to one formulated by Marx (9), although Marx abstained from applying this idea to science. Sohn-Rethel, however, has made a step in this direction (10).

But it is not just parsimony or economy that moves ruling thought. There exist other preconceptions, which show their influence when we arrange our data, the results of extrasubjective sense experiences in categories.

Categories are more often than not the outgrowth of social activities and of unscientific thought (11). It is these unscientific preconceptions which then mould the arrangement of data. I have no objection to this, except the one that only certain preconceptions, those enshrined in a given educational system are involved. To give flesh to these bones, I would like to exemplify this.

No single good scientific reason exists to think of our sun as a material body in space. Even less good reason — save the parsimonious one — supports the claim for a central position of such a radiating non-object. The Pythagorians put fire at the centre of the universe, in reminiscence of the then social centre, the fireplace or hearth. Copernicus, much reputed forefather of our scientific age, put the sun into the centre of the universe. Surprisingly little is said about this move. What moved Copernicus is never asked, taking it for granted that he was in insatiable love with scientific truth. Our rational minds like to overlook the fact that Copernicus was as much an astrologer as others of his contemporaries, Kepler, Tycho de Brahe, etc. An ancient association exists in astrology between planets and metals. The sun, also taken as a planet in astrology, is affiliated with gold. The fact is too often overlooked that Copernicus was deeply interested in economics, in particular in inflation and the value of money. The Copernican Law (13) of economics is hardly mentioned when talking about the heliocentric system and Copernicus' suggested remedy for inflation, a strong central authority (the later absolute 'sun king') controlling the gold standards of currency, is ignored. The concurrence of a central authority controlling gold with a golden planet put into the centre of the universe is good for thought. I would say Copernicus put economics into heaven, together with an absolute king, thus projecting a desirable social arrangement into heaven, which a little later Hobbes announced more outspokenly.

It might be counter-argued that the source of inspiration is irrelevant as long as the derived scientific facts are proven true. But truth is always to be negotiated. Copernicus' deep involvement in mathematics derives from this need for arguments in negotiations, his longish treatment of other possible counter-arguments show the same concern. The weak position of Copernicus' counter-arguments was even clear to Galileo (14). Truth is negotiated and it is never clear when a debate will be exhausted. Quite recently, when Michelson's experiments failed, a geocentric system was again debated and reconsidered among physicists. Then Einstein's theory of relativity was introduced, and from this point onwards the central position of the golden sun became relativized too, as everything moves. Whether relativity should be understood as a projection of the collapse of a substantial number of monarchies during the first decades of this century is a matter I have not had the opportunity to study in depth. I do presume, however, that as a consequence of relativity theory and ongoing social developments, it is to be expected that the image of our sun will change in the future, possibly quite radically. The preconditions for such a switch are given and I will not hesitate to sketch mockingly such an alternative later on.

Sun and planets aside, similar examples can be found in other contexts where social experiences of some kind become transplanted into quite different subject areas. Sohn-Rethel argues that our conceptions of time and space are the result of exchange relations, where, through application of the 'Tauschpostulat' (15), time and space are stripped from any qualities of their own, to keep the ones of commodities constant in principle. Toulmin and Goodfield, although firmly arguing along the line of prevailing science-ideology, provide further material, when saying, e.g.

... nor did they find it difficult to imagine (e.g.) one hundred thousand pounds in money: why, then should their minds recoil from seventy or hundred thousand *years*? (16)

To these authors it is clear beyond doubt that the geological findings of the 18th and 19th century together with social imagery drawn from economics made it possible to push the limits of history beyond the given biblical 6000 years. Sense data alone would allow for many alternative interpretations, discontinuous jumps, as Cuvier's catastrophy theory suggests, cyclical arrangements, as Plato's great year would make plausible, and last but not least linear arrangements, as we accept today, with its *necessary* expansion into billions of years, while knowing of billions of dollars in the hands of some firms and states.

None of the possible alternatives was refuted; it pleases us to arrange our data in a linear manner when moving in some other contexts. It seems therefore not unlikely that Feyerabend is right when he argues that disreputable theories or sciences might come back one day (17). Sufficient examples of comebacks exist, be it money (18) or atomism, be it fashions or the recent re-emergence of the so-called 'pseudosciences'. The arrangement of data is a question of certain preferences, which themselves depend on social and educational factors.

It should be stressed at this point that the arbitrariness of arranging data is by no means restricted to the historical sciences. René Thom, for instance, argues convincingly (19) that in applied system theory scatter diagrams are too quickly transformed into some smooth statistical function, interpreting irregularities as 'noise in measurement'. Thom takes such data more seriously, ingeniously pulling some of his catastrophes out of the dots, while others find nothing but linear and quadratic functions. It must be clear that we attribute to fundamental qualities such as time, space and mass and data properties at will or refuse them at will.

This alone, however, does not constitute the whole of the story. Not only do we allocate properties at will, in addition we also give preference to certain activities and operations at the expense of others. Take one of the most fundamental statements; $1+1=2$. This statement is not infrequently used to convince heretics like myself of the indisputable existence of absolute truth and, by applying it, to make us shut up. And indeed more often than not the desired result is obtained. The method applied is simple, it is a belated appeal to the authority of school.

To formulate a counter-argument one must forget the teachings of teachers and start all over again. The notation $1+1$ denotes an operation of the following kind; it means: take one object, place next to it another object, the result of the operation is two objects, hence 1+1=2. As long as people restrict their attention to this kind of operation the statement is indisputable, 1+1=2. But there exist other operations, for instance, the following one: Take a strip of paper, glue to its end another strip of paper; the result is one object, one strip of paper, hence 1+1=1. Ah, possibly the protagonist of absolute truth will claim, the first strip had the length of one metre, the second had the length of one metre and the fusion of the strips has, you can easily check, the length of two metres. Consequently 1+1=2. We obviously performed but one operation and *achieved two results*. We have two methods of representing the operation. Both arguments are correct, the only difference being that the protagonist *reduced a complex reality to one single aspect*, to the activity of merchants who put a metre at the end of the other to calculate the value of their merchandise. Out of a multitude of possible operations one was singled out and given preference in congruence to prior socialisation and education. This concentration of attention on a single aspect must have unexpected consequences, when entire nations succumb to the habit (20). The restricted frame of perception in mathematics to some basic operations is certainly one of the reasons why mathematics has so little to offer to sciences where fusion and synthesis are of prevailing importance, amongst them chemistry, biology and the social sciences, except economics.

The list of most preferred operations can be prolonged. Another item in the list is the operation 'cutting out' or 'dissecting'. Analysis is precisely this operation of cutting out. It constitutes one of the methodological virtues of present day science. Mostly it is intertwined with another performance called experiment. The sequel of operations reads in plain words: Cut the matter out and bring it to safety, out of danger, ex-pericolo. Analysis and experiment are complementary actions, like analysis and addition. Only after having isolated an object from its context it can be put next to another object and be added. Since analysis forms a fundamental concept in science, science can be understood as the art of reducing the complex web of nature through 'incarceration' (21) to a point where a phenomenon becomes so reduced through isolation that it can be manipulated at will, just like people after prolonged isolation. 'How to manipulate?' seems to be a principal, triggering question in science. It is reminiscent of its origin where the *hand, manus*, of the artisan was the performer of experiments directed in its activities by the *will for production*.

It is at this point that the perspectives of the two bourgeois companions, the merchant and the artisan, fuse (22). It is the interest of the artisan to cut out phenomena from their natural context to subject them to manipulation, and it is the merchant's interest to cut products out of their social context of origin to subject them to other hands, to his 'Handel' (23). Mathematics was for long nurtured in the domain of commerce, hence the registered disregard for activities of fusion in this discipline. Fusion is not part of the cognitive world of exchanging merchants.

Reducing principles of science to basic experiences of some social groups triggers the question of the possible existence of alternative sciences with different codes of conduct and different knowledges. The term 'science' is of recent European origin. It would be but fair to apply it exclusively in this historical context. Seen from this perspective no true science existed elsewhere, either in China or in antiquity. One might, however, also agree to call scientific any social institution producing knowledge and world images on the basis of some agreed code of conduct; then science is indeed a general social phenomenon to be found in most societies. I personally give preference to the last version, a step inviting us to compare various sciences, their results, methods and social roots. Indulging in this enterprise involves more than just one research project and consequently surmounts the scope of this essay. For consolation we will be moderate and satisfy ourselves with a few rough sketches.

I draw this example from the European middle ages, not least because many people refuse the idea of the existence of a medieval science. The refusal is valid as long as we commit ourselves to the narrow definition of the term science. The broader definition makes, however, the idea of a medieval science perfectly acceptable. As proof of this, the first European universities, were the medieval institutions for producing, disseminating and dismantling knowledge. True, their central concerns were of a different nature then those of our present science, but both produce valid world images.

The world images produced then and those produced now have little in common. Methods, however, are not at all that varied. Medieval scientists were well acquainted with the notions of tests, experiments and emotionless investigation, although they took a different form. Emotionless investigation was practiced during the inquisition (24) and tests and experiments had the form of divine interventions through miracles. Their main outlook was oriented towards exceptional and extraordinary events, and were the least concerned with production of goods. 'How to do things', 'How to manipulate things' was not at the core of their interest. Medieval science was the science of an aristocratic society, it was done in the interest of an elite of warriors. Theirs was a different world view. Their central concern was that of securing sufficient followers (25) by distributing booty and glory in a justifiable manner. Theirs was consequently a science of justification and legitimization, law and theology. Religion formed the ultimate principle which put an end to infinite bargaining and negotiations, not dissimilar to scientific expertise nowadays.

Warrior aristocrats have no interest in production, nor even in possessions and life. Their dominant interests lie in social relations and glory (26). Life experiences of warriors have little or nothing in common with those of artisans or traders. The successes of these are cumulative and repetitive in nature. A single sale counts little, their fortune has to be accumulated, whereas the fortune of warriors is instantaneous and non-repetitive. Dying is a single event, but crucial, rupture is a single event, defeat is a single event. Making decisions on the basis of great numbers of events is ludicrous when confronted with those alternatives. How different the perspectives of artisans and traders! A single sale hardly keeps production going; theirs are great numbers, theirs are repetitive and predictable events. Those differences in perspective should throw light on the differences in proof; here the exceptional miracle, there the repetitive output of a controlled production machine (27). It is indeed frequently said that science is knowledge about great numbers, once is nonce (28). Aphoristically one might say, our science is science looking for the largest common denominator, whereas medieval science sought the smallest common multiple.

The observed tendency for abstraction in our science is most probably one result of this difference in outlook. It creates, when applied to human affairs, alienation in form of the many 'homo'-images, be they politicus, oeconomicus or ludens (29).

It should have become evident by now, that not only our images of space, time and matter are intrinsically formed by the dominance of the economy of commodity exchange and production, but the image of ourselves, as presented through the various sciences of psychology, biology, history and other humanities, including sociology, this alienated little passport photograph that we carry around in our heads, is a snap-shot taken through the lens of this prevailing mode of thought. How this method of construction of 'homines' through abstraction from reality functions together with its consequences is shown in detail by Sève (30).

Little wonder that between the world of abstract beings and of real beings there is an irreconcilable gulf, where discontent and disillusionment grow and where the seeds of counter-sciences happily flourish (31). This gulf between the necessities and comforts of the layman and the idealized constructions through abstraction and experimentation takes many forms. It takes the form of heart transplantations on the one side and of an increasing poverty of medical diagnosis on the other; of space shuttles here and of the breakdown of public transport there; of economic prognosis here and of unemployment and inflation there; of chemical fertilizers here and of waste and pollution there. The list may be prolonged at will. The scientists' usual defence is well known, theirs is not the application of knowledge and consequently they are without responsibility. Notwithstanding their good intentions, their apologia is in vain. Their myopia in procedure frees them from murder, not from manslaughter.

The verdict might sound harsh, but eminent scientists themselves, amongst them Shapiro or Russell (32) consider the continuation of our science as criminally irresponsible. 'All modern scientific thinking is at bottom power thinking' reads Bertrand Russell's judgement. These and other scientists think that the practice of science is, in principle, hostile to life, we would say because of its repetitive and exclusive concern for great numbers and for regularities with the consequent appraisal of a static, lawlike and immutable world. One might twist Goethe's self-introduction of Mephistopheles around by saying science is 'ein Teil von jener Kraft, die stets das Gute will und stets das Böse schafft' (33).

At this point, I imagine hearing an outcry of protest from some of my readers. Was not science the unique, liberating force, that brought longevity, liberated humanity from epidemic diseases, superstition and paucity? Was life not nasty, brute and short before science's victory?

Replying to such sweeping statements is difficult. The answers definitely need more care and elaboration than exhibited by the crude generalisations made. On their extremely broad and undifferentiated level the above assertions are definitely wrong. Not even during the Stone Age was life only nasty, brute and short, as Sahlins (34) shows convincingly. McKeown's (35) analysis of the decline of death rates demonstrates that the disappearance of the great plagues of mankind are, in most cases, not attributable to scientific advances in medicine, but rather to improvements in agriculture and nutrition. It should be remembered that, in the 17th and 18th century, agriculture was no scientific discipline like physics. It is also a simple fact that the black death disappeared from Western Europe in the 17th century, from central Europe by mid 18th century and from all of Europe by 1841, the date of the last epidemic in Archangelsk. Mumford's (36) presentation of the 'eotechnic period' during the European middle ages also paints a radically different version of this period than the one to which we are accustomed; namely, as being dark, brutish and dull. Furthermore, there is the evidence of anthropological findings, such as the North American Indians concept of the normal length of human life being 70 years, the practices of Indian Brahmins of smallpox vaccination or the complex surgeries performed by the Ashanti (37).

It is, however, not the intention of this essay to belittle the positive results of science. Great advances are to be registered, even if a fair number of them, (for instance, penicillin) are less the result of scientific discipline than the findings of an open mind, registering carefully the activities of an immediate environment (38). Despite of my sentiments and hesitations, I wish to stress at this point, that science was indeed a liberating force. It was a method that freed Europe from the dominant system of thought of a dominant, rotten class. It must, however, be kept in mind that it was not the sole and only liberating force in history. Christianity was a liberating force, too, in its early days, and remained so until it became intermingled with power (39). In its early days Christianity was no science and had no science. Later, however, it gave birth to theology and from then onwards it rapidly ceased to be liberatory and became instead an opiate for the people, a good source of income, an instrument of control and a social remedy for the then ruling classes.

As an analogy ot this historical development one might say science, too, was a liberatory force before it became intermingled with power. From this point onwards, it deteriorated, turning into a new type of opium for the people and a good source of income, etc. This analogy between science and Christianity is tempting. Both come from humble origins. Both fought a seemingly unequal combat with the establishment. Both made their historic compromise with power. One of them was corrupted by this move, turning into an official institution for producing official knowledge and world images. The other is on the brink of petrifying into this state — if it is not petrified yet (40).

Feyerabend (41) requests the separation of science from the state, just as church and state should be separate. In all likelihood, this would mean no big money for Big Science, no sky rockets, no accelerators, no research centres for cancer, no biology labs for gene manipulation, only Little Science at the kitchen sink, without shockproof concrete walls, without gigantic telescopes and electron microscopes. What a retreat and what a relief.

This is, however, most unlikely to happen. The inducements for scientists are too sweet, the pay-offs for power too great. Scientific *facts are facts*. They are facts not to be negotiated, it is said; no opinion poll is to be taken, no vote exists on which these facts depend, it is said. Little surprise, therefore, that state and power takes eminent interest in science and its protagonists, in the scientific elite of experts for absolute truth (42). Just as a theological verdict supported monarchical claims, so does scientific expertise create what our ruling class needs to remain in power, the so-called objective establishment of factual constraints (Sachzwang) of extrasubjective facts. The presumed despotism of nature and of natural laws rectifies the one of potential despots and laws.

Again, it is to be expected that some voices will be raised in protest, as ours is a pluralistic democracy. Democracies are social entities, where, as everyone knows by heart, each person is free to think and to say what they believe to be right. One restriction clearly exists, however. You, the non-expert, must not raise your voice against the superior knowledge of experts, usually scientists. The undeniable advance of science in every field and its intrusion into every field consequently reduces the scope and breadth of fields where non-experts will be able to say and to do what they believe to be correct, in a reciprocal manner. The more science advances the less room will be left for liberty in speech and action. This constitutes a strange concurrence, as science sprang, as we have already said, from a democratic movement and was once indeed a liberating force.

The development of science is a dialectical process, as every historical process is.

Even under the overoptimistic assumption that no politicians wish to manipulate themselves into positions of unrestricted power, the existence of indisputable knowledge (43) — indisputable by laypeople (44) — constitutes a deep threat to democratic principles. When nothing can be disputed — except by some experts — democracy vanishes. This change in the nature of science transmutes science into its opposite and robs it of its liberatory cloak, scientific power thinking turns into power science. The possibility of this development was already sensed by Bernal (45). In his controversial future world, a small layer of experts keeps the rest of humanity lulled into pleasantries. Social development seems to tend into the direction of the technocratic rule of scientific managers (46). This situation will change science fundamentally. It will cease to be a liberatory force, just as Christianity ceased to be a liberatory force in history.

Christianity frequently changed its costumes in the course of time. The religion of St. Augustine was different to the religion of bishop Arius and different to that of the Carolingian Empire. The theology of Albertus Magnus was not the same as the theology of the Aquinate, which differs again from Eck's, not to speak of the various forms of protestantism. In a similar vain, science changes its faces and its emphasis. Science at the turn of this century will not be the same as science at the turn of the next. Not only will the corpus of knowledge change, also scientific disciplines themselves will change, as the dominant questions exert their impact on them and those will alter with the dominant classes.

If the past was the age of merchants and artisans, and their hybrid, capitalists, so the future is most likely to be that of scientific managers. Not 'How to produce things with profit' but 'How to manage things with profit?' will be at the basis of scientific interests. Not only will agreement on methods change in the sequel of this basic switch in interests, but also the offered world images will be very different. At the core of interest then there is no longer matter, but information. The human body will not be seen predominantly as a material body (47), but as a complex system of flows of information, comparable to a computer and to be analyzed and represented like a computer program.

The switch in basic outlooks will have its effects on everything. The image of the cosmos will change. Perhaps it will turn into a four-dimensional universe, where the fourth dimension is not time, as in relativity theory, but also space. Our own universe of three dimensions is then but a subuniverse of the other in which four-dimensional substance exists, an integrated form of matter, energy and information. Four-dimensional substance pours into our own universe, where it will be transformed into energy, matter and order(s). Matter and energy flow back into four-dimensional space through sinks, possibly the present black holes. The sun, always important in an inegalitarian society, will be thought as a 'source' where four-dimensional substance hits the three-dimensional world of ours (48). In our world of three dimensions a class of engineers controls and commands, a class of scientists of leisure contemplates and a class of 'hands' wangles and wrangles for three-dimensional matter — products. The four-dimensional universe is not accessible to ordinary beings, in it resides the great organizer and planner — alias god-father — in a well-tailored suit with a huge computer at his disposal. And god-manager loves manageology and similar disciplines (49).

It is to be expected that some readers might attribute to these ideas some definite degree of crankiness. How do I come to make such prophecies from which even not very scrupulous system-theorists or econometricians will shudder back? My reply is simple. First, I am no system-theorist (50) and, second, I would say that the social development which is the basis for the new images seems to have been on the march since decades, at least since the days of Veblen and Weber (51). When bureaucracy and conspicuous consumption become an integral part of economy, then vast branches of industry turn into industries producing symbols. Use value changes its nature, as the predominant use of goods changes radically. The commodities of our age become increasingly those of display. Weaponry, as much as means of transport, or even capital goods exhibit this tendency. Through this production of symbols turns gradually into symbolic production, where hard material products become an insubstantial 'appendix' of contracts (52). The commodities of the future exhibit increasingly the characteristics of symbols, signals, fashions, blueprints of complex networks, pictures and hallucinations. The most desirable good to be produced is then desire, the offspring of demand.

However, there are also limits to this development. At the point where marketing and advertising strategies find it increasingly difficult to succeed, crisis management will have to step in. But crisis management will not restrict its activities to 'natural crisis', but expand as a science to the same degree as physics did, i.e. not stop still at the point of description of a 'natural environment'. Just as physics expanded into an artificial world of low temperatures, high energy, and fast particles, so will crisis management expand into a world of 'artificial crisis', producing and controlling them at will.

And science will not stand still at this point. It will not satisfy itself with changing the social and political world, it will alter science itself (53). When the dominant concerns of a society become managerial, because its dominant class is managerial, then the shape and products of science undergo metamorphosis and become comparable to medieval science (54). Boundaries between disciplines will change as data is grouped together in novel ways, bringing new disciplines to the fore and making others disappear. Themata and world images will change. The presentation of an argument will be more important than its content (55). Computer science, semiology, semantics (56), cybernetics, system theory, operation research, artificial intelligence, networks, recruit the dominant disciplines. Old-fashioned physics might suffer a similar fate to the old-fashioned astrology of some centuries ago. It is not unlikely that this ancient 'queen of the sciences' will raise its head again celebrating its comeback in a new robe, accompanied by telepathy, and a new version of theology.

In this world of masterminds, however, the substantial necessities of ordinary people will be crassly neglected. This is the source for fresh underground movements, for knowledges about small-scale material production, from which, in a distant future, some new science might emerge, if the movement should prove successful. The ruling classes of this age will react in a comparable manner as our own do. Ridicule the primitive enterprises, make their own scientists measure away the deficiencies of their power, mystify the neglected needs by talking about 'Homo's' primitive animalisms, all this will be part of the tactics of power science. The superstition of the age will be the belief in 'doing', as 'doing' was a basic concept in the dark and smoky age of our days. The frustrated lower classes will learn to disregard the promises of their unfulfillable life-accounts on computers and start to pass knowledge of cooking, reading, writing, shitting and breeding secretly from friends to friends. This constitutes the other side of Bernal's vision of brains in silver containers. It is the perspective of the forgotten part of society, the never negligible portion of people for whom scientific progress means predominantly regress, as no science has, until the present day, succeeded in being progressive for all. Not in science is progress to be found, but in the social means of distribution. The advances of science mean little, if they are not equally distributed.

At this point it might be appropriate to come back to my introductory remarks. Science, as I understand it — be it medieval, present or future — is monomaniac in its methods (57), and despotic in its performance. Science is usually deeply involved with political power. If it is not, then it ceases to be science, falling back into folk knowledge, which is inconsistent but favourable to the needs of people, including the need of being free to arrange data and sense experiences according to one's own tastes.

The claim for an all embracing cognitive world is a power claim. Proclaiming the separation of cognitions and of values is a lie (58). Each concept, each category is rooted in a previous value judgement. Science selects *and* proves its phenomena under experimental conditions. Its insights are consequently by necessity circular and constructivist. Scientific knowledge is, even in relation to this fabricated reality, at the best approximate, a fact that application of this knowledge in any form of technology proves with ample evidence. My own little games aim for a different kind of reality, as reality is always construed. The difference is lack of power, or gain in equality. The more people play their own little games, the more would we approach a state of affairs which resembles egalitarian and humane science, at best, and beyond this it would help to create a solidarian society. In this situation neither paradise nor hell will result. The devil in his many forms will still be at work. This devil will be human in scale and much smaller than under the rule of some paradisical partriarch or matriarch.

Acknowledgements

I am indebted to Ingo Grabner and Helga Nowotny for many helpful comments, and to a square Vienna Circle of critical irrationalists, who encouraged me to put into writing what we thought jointly in the course of a year's weekly sessions.

Part of this essay was conceived in a long conversation with Steven Rose, whose views on science can only be guessed by my own one-sided replique to them.

Notes and References

1. It is well known that the law of gravity has to be adjusted by the use of the 'Reynold number', when non-experimental conditions are given, this means under reality. Further: only the rare, inert gases react according to the principles of thermodynamics. When applying these principles to other gases, particularly steam, they have again to be adjusted.

For combustion engines see also G. Küppers, 'On the Relation between Technology and Science'. In W. Krohn, E. Layton, and P. Weingart (Eds.), *The Dynamics of Science and Technology, Sociology of the Sciences*, Vol. II, D. Reidel, Dordrecht, Holland, 1978.
2. P. Feyerabend, *Against Method*, NLB, Humanities Press, London, 1975.
3. E. Schrödinger, *Was ist ein Naturgesetz?* R. Oldenbourg, München-Wien, 1962.
4. *Ibid.*
5. N. Georgescu-Roegen, *The Entropy Law and the Economic Process*, Harvard University Press, Camb., Mass., 1971.
6. It is said that relativity theory and quantum theory reduce the number of possible mental constructions more or less to one. It is a strange argument, as we have to accept first a number of assumptions, starting with the constancy of the velocity of light, leading to experimental arrangements, i.e. we have to restrict our own perception first, before we reach this extremely restricted world view. What happens if we decide not to accept the request?
7. A nice example for the various ways of interpreting data may be found in the context of empirical political science in A. Przeworski and G.A.D. Soares, 'Theories in Search of a Curve: A Contextual Interpretation of Left Vote', *The Am. Pol. Science Review*, LXV, pp. 51–68.

   Interpretative schemes depend on cultural aspects of epistemologies. See on this, e.g., M. Maruyama, 'Heterogenetics and Morphogenetics; Toward a New Concept of the Scientific', *Theory and Society*, V/1, 1978, pp. 75–95.
8. Although it is true that an individual firm tries to be 'economical' when calculating its total costs of production, it is certainly not 'economical' on a national level, when trying to maximize its sales. The literature on this topic is legion, think of J.K. Galbraith, *The Affluent Society*, 1958, to mention but one.

   Some scientists demand 'beauty' in their theories. The roots of beauty are, however, even more deeply sunk in preconceptions of social constraints. See P. Bourdieu, *Zur Soziologie der symbolischen Formen*, Suhrkamp Taschenbuch StW 107, Frankfurt a.M., 1974, to mention just one prominent representative.
9. K. Marx, 'Die Deutsche Ideologie'. In S. Landshut (Ed.), *Die Frühschriften*, A. Körner Verlag, Stuttgart, 1964.
10. A. Sohn-Rethel, *Geistige und körperliche Arbeit*, Suhrkamp Taschenbuch 555, Frankfurt a.M., 1970.
11. It is, of course, also true that scientists fight back. Our own thought becomes increasingly more scientific. No one, who wishes to be educated, would use the term 'natural element' any longer when referring to fire, water, earth or air. Similar examples exist in great numbers.
12. Since it is a custom in science to add weight to one's argument through referring to authority, particularly Nobel Prize winners, I quote from E. Schrödinger:

    Das tiefe Erdinnere, das Innere der Sonne und der Sterne gehören hier her; ja schon daß Sonne, Mond und Sterne in Entfernungen, die sich mit der Meßkette nicht nachprüfen lassen, *als materielle Körper* (my emphasis) greifbar im Raum schweben und etwa einem aufprallenden Meteor Widerstand entgegensetzen würden, sind unbeobachtbare Züge des Weltbildes, die wir doch daraus nicht fortlassen können oder *wollen*.' (my emphasis). E. Schrödinger, *loc. cit.*, pp. 47–48.

I translate:
> The deep interior of the earth, the interior of the sun and of stars belong to here; yes, even that sun, moon and stars swim in distances, which cannot be checked by a surveyor's chain, *tangibly as material bodies* in space (my emphasis) and possibly put up resistance to a bouncing meteor, are unobservable features of our world image, features which we cannot or *do not want* to dismiss (my emphasis).

13. The negligence is comprehensible, as we have here one other example of narrow chauvinism. The tendency that bad money drives good money out of circulation, was rediscovered by Sir Thomas Gresham, and is consequently called 'Gresham's Law' in Anglo-Saxon countries.

    Some might argue, that Copernicus wrote his *Commentariolus* by 1512, where he already sketched the heliocentric system roughly. His economic writings are dated 1519, 1522 and 1528. Consequently one could argue that he put heaven into economics. I am not the least impressed by this argument. The 'Little Commentary' is a pamphlet which provides an outline for a system of thought. We do not know what moved Copernicus at the age of twenty or thirty. What we know is that e.g. Neoplatonism existed in those days and that scholars like Marsilio Ficino were known at Krakow University, together with the emphasis put on the central and divine position of the sun. Arguing along these lines would only move our central concern further back in history, as we had to ask then, what moved the Neoplatonians?

14. See P. Feyerabend, *Against Method*, Chapters 6–9, and Appendices 1 and 2, *loc. cit.*
15. Tauschpostulat (postulate of exchange) demands that a good must not change its qualities while passing from the ownership of one person into that of another. If transport is necessary during this period, time and space must not exert any influence on the good. Consequently, time and space are understood as having no qualities of their own.

    See A. Sohn-Rethel, 'Geistige und körperliche Arbeit', *loc. cit.*
16. S. Toulmin and J. Goodfield, *The Discovery of Time*, Penguin Books, Middlesex, 1967, p. 180.
17. P. Feberabend, 'Unterwegs zu einer dadaistischen Erkenntnistheorie'. In *Unter dem Pflaster liegt der Strand*, Vol. IV, K. Kramer Verlag, Berlin, 1977.
18. It is a historical fact that during the early Middle Ages a severe shortage of currency existed. Many people say, therefore, that medieval economy was a natural economy. Scholars, like M. Bloch, weaken such statements to saying:
    > Not only was money generally scarce, and inconvenient on account of its unreliability, but was circulated too slowly and too irregularly for people even to feel certain of being able to produce it in case of need. M. Bloch, *Feudal Society*, Vol. I, Routledge and Kegan, London, 1971, p. 67.

    On the possible *re*-disappearance of money in circulation see e.g. S. Bodington, *Computers and Socialism*, Chapter V, Spokesman Books, Nottingham, 1973.
19. René Thom, from a lecture delivered at the 4th European Meeting on Cybernetics and system Research, University of Linz, Austria, March 28–31, 1978.
20. The social rooting of mathematical principles was clearly seen by L. Wittgenstein. See on this subject D.L. Phillips, *Wittgenstein and Scientific Knowledge – A Sociological Perspective*, Chapter VI, Macmillan Press, London, 1977.

21. 'Incarceration' means making a prison around a thing or phenomenon. It differs from 'imprisonment' insofar as the thing is not put into prison, but the prison is made around the thing. The effects are the same, but the approach is different. To proceed in the opposite manner frequently does not bring the required success. The entire procedure can be compared to catching animals by traps. Putting the trap in the wrong place usually brings little success. This can be illustrated by the following quotation from A.L. Sengers et. al.:
    > Near a critical point the behaviour of widely diverse systems simplifies to one of a few universal patterns — but fluids for years resisted *being fit into the slot* (my emphasis) that theory had prepared for them.

    In 'Critical-Point Universality and Fluids', *Physics Today*, (12) (1977) p. 42.

    It should be stressed finally that from the method of 'catching' no conclusion can be drawn to what procedures the captive is subdued afterwards. It is not defined at this point whether the animal is made into a stew or made to jump in a circus. The only thing sure is that the behaviour of the captive in all likelihood will change quite radically.
22. The social origin of science is well portrayed in E. Zilsel, *Die sozialen Ursprünge der neuzeitlichen Wissenschaft*, W. Krohn (Ed.), Suhrkamp Taschenbuch STW 152, Frankfurt a.M., 1976.

    One will also find there interesting observations about the interrelations of commerce and the revival of mathematics.
23. 'Handel' in German means commerce. The word is reminiscent of the ancient relation between changes in ownership and changing hands. The English words 'handsel', 'hansel' and 'handle' come from the same etymological root. See *The Oxford Dictionary of English Etymology*, Clarendon Press, London, 1966.
24. Contrary to erroneous, widespread beliefs the Inquisition is much older than thought. It can be traced back to 1183 as an outgrowth of a pact between Pope Lucius III and Frederick I in Verona.

    See F. Heer, *The Intellectual History of Europe*, Vol. I, p. 162, Anchor books, Doubleday, N.Y., 1968; W. Ullmann, *The Individual and Society in the Middle Ages*, J. Hopkins Press, Baltimore, 1966.
25. This is a common feature not just restricted to Europe, see F.G. Bailey, *Stratagems and Spoils*, Chapters 3 and 4, Basil Blackwell, Oxford, 1970.
26. See N. Elias, *Die höfische Gesellschaft*, H. Luchterhand, Darmstadt, 1975. N. Elias, *Über den Prozeß der Zivilisation*, Suhrkamp, stw 158, Frankfurt a.M., 1976.
27. The foundation of rules in practice, experience and action forms an important topic in L. Wittgenstein's *Philosophical Investigations* I/202, 206, etc. See also N. Rescher, *Primacy of Practice*, Basil Blackwell, Oxford, 1973.
28. The significance given to this rule becomes overt, when looking at teaching material. F.R. Bradbury writes e.g.
    > ... I would enshrine in the ground rules of scientific method ... "einmal ist keinmal" (once is nonce).

    In F.R. Bradbury (Ed.), *Words and Numbers — A Student's Guide to Intellectual Methods*, p. 74, Edinburgh University Press, Edinburgh, 1969.
29. For a quarter of a century another Homo has haunted the minds of sociologists, Homo Sociologicus, a being connected with the name of R. Dahrendorf. The little booklet with the name of this fantastic being on its cover is good demonstration of

what happens when reducing people to 'Homo' of some kind. It is surprising to see how an educated person can think and write so much about roles and stages, without discussing the playwriters' 'role' in the process. As a result of this abstract procedure, abstract concepts like 'free will' and related concepts are more easily at hand than economic relations and/or relations of dominance. The resignative final statement is, however, worth quoting, even though I do not fully agree with it, it reads:

> *Images of man*, too, *belong in that sphere of paratheory* (my emphasis) where the most fruitful stimuli for new theories may originate, even if the stimuli themselves are personal rather than testable by experience.

In R. Dahrendorf, *Homo Sociologicus*, Routledge and Kegan Paul, London, 1973, p. 88.

30. L. Sève, *Marxisme et théorie de la personalité*, Paris, 1972. German and English translations exist.
31. The reader will not be surprised if I tell him/her now that the re-occurrence of counter-scientific movements is readily explained by another feature of 'homo's' abstract character. It is 'homo's' deep rooted desire for 'irrationalities', which science does not satisfy accordingly. Hence the new wave of superstitions and the need for rediscovering legitimate faith. Consequently, writes H. von Ditfurth,

> ... Aberglaube in der Vielfalt seiner Formen ist es heute vor allem, der die Chance zur Wiederentdeckung *legitimer Glaubensmöglichkeiten* (my emphasis) unter sich zu begraben droht.'

In *Der Spiegel*, XXXII/17, 1978, p. 55.

> ... superstition in its many forms threatens to bury the chance of rediscovering the chance for *legitimate forms of faith*. (my emphasis))

One man — homo —, one truth, one faith, one law!

32. As quoted in B. Easlea, *Liberation and the Aims of Science*, Sussex University Press, London, 1973, pp. 251, 264, 328.
33. The correct quotation reads: '... ein Teil von jener Kraft, die stets das Böse will und stets das Gute schafft.' J.W. Goethe, *Faust*, Part I, Studierzimmer.

    I translate *my version* of the quotation: '... part of the force that always aims for the good, but gives continuously birth to the bad.'
34. M. Sahlins, *Stone Age Economics*, Tavistock Publ., London, 1974.
35. T. McKeown, *The Contribution of Economic and Social Influence to Improvement in Health*, Con. Int. su Medicina, Economica e Societa nell' Esperienza Storica, Pavia, Sept. 27—29th, 1973.
36. L. Mumford, *Technics and Civilisation*, Harcourt, Brace and Co., N.Y., 1934.
37. These examples show that it is much less the amount of knowledge available than the distribution of knowledge and of resources which liberate society. Knowing about smallpox vaccination or about heart transplanations alone does not liberate society, on the other hand, if economic and social conditions are healthy themselves, then scientific knowledge contributes only marginally to the well-being of a society. This is one consequence of McKeown's research. Or to give another example, Maria Theresia, empress of the Austrian monarchy for 40 years, gave birth to 16 children without being physically hampered in governing a complex empire. It was the accompanying conditions of her life which enabled her to do this, not science. Maybe I should add that I do not argue for any increase of birth rates.

38. It is commonplace to say that creative insights have nothing to do with science, they are independent of methods. See on this, for example, Einstein's position as represented in G. Holton, *The Scientific Imagination*, Chapter III, Cambridge University Press, London, 1978.
39. That religion did not cease to be occasionally a liberating force even in recent history becomes clear when reading S. Yeo, 'A New Life: The Religion of Socialism in Britain, 1883–1896'; *History Workshop*, Autumn 1977/4, pp. 5–56.
40. This analogous development of religion and of science is also seen by others. E. Mendelsohn writes, for example, of scientists as emerging as a new priesthood in alliance with the state. See E. Mendelsohn, 'Should Science Survive its Success'. In R.S. Cohen, J.J. Stachel, and M.W. Wartofsky, (Eds.), *For Dirk Struik – Scientific, Historical and Political Essays in Honour of Dirk Struik*, Boston Studies in Philos. of Science, Vol. XV, D. Reidel, Publ. Comp., Dordrecht – Boston, 1974, pp. 373ff.
41. P. Feyerabend, *Unterwegs zu einer dadaistischen Erkenntnistheorie, loc. cit.*
42. For current developments, see H. Brooks, 'The Federal Government and the Autonomy of Scholarship'. In C. Frankel (Ed.), *Controversies and Decisions*, Russell Sage Fdt., N,Y., 1976, pp. 235–258.
43. This is neither to say that scientific theories are not disputed within science, nor that hired scientific experts do not promote controversial evidence. Every political group pulls its scientific experts out of its sleeve. This phenomenon is not only connected with the debate about atomic power plants, but constitutes a historical fact, again since the days of medieval disputes about taking interest on loans. It means, however, that non-experts are excluded, as a rule.

    See on this topic, H. Nowotny, 'Controversies in Science', *Zeitschrift für Soziologie*, IV/1, 1975, pp. 34–45; H. Nowotny, 'The Social Aspects of the Nuclear Power Controversy', RM–76–33, IIASA, Laxenburg, 1976.
44. Lay are sometimes invited to a spectacle, where a series of scientific experts pull an even larger series of rabbits out of their bowler hats. But as it frequently happens in magic, it is very difficult to say which of the white rabbits are red herrings.
45. J.D. Bernal, *The World, The Flesh, and The Devil*, Indiana University Press, Bloomington, 1969.
46. A classic on this topic is, of course, J. Ellul, *The Technological Society*, Knopf, New York, 1964.

    But see also H. Skolimowski, 'Does Science Control People or Do People control Science'. In R. Clarke (Ed.), *Notes for the Future – An Alternative History of the Past Decade*, Thames and Hudson, London, 1975, pp. 130–138. *But* – the entire book is very enjoyable to read.
47. This development has again been on its way for some time. An example for the direction of the course can be found when looking through the programmes of recent conferences. I want to list just a few headings from the 'Symposium on Control Mechanisms in Bio- and Eco-Systems', Leipzig, Sept. 1977. Among the topics are: Control within the organism, Sensomotor control mechanisms, Autonomous control of organ functions, Control of growth, Control of immune processes, Control of cell metabolism, Communication between organisms, Control: Organism/environment etc. Control looms large, wherever you look.
48. I want to stress that what I say here is not just pure fantasy. In January 1976,

various independent sources announced the observation of solar pulsations which not only change the shape of the sun at regular intervals of 2h 40 min., but also put into question current theories about the origin of the sun's energy. The responses to this 'scientific crisis' include the suggestion that the energy source might be a 'black hole' in the centre of the sun. Another suggestion says that solar energy comes from a process unknown to science. As you see the sun opens up and begs for integration into another world perspective. The linkage to black holes exists in cognition already, you will be 'surprised' to hear one day that the link exists in 'reality'.

See J.N. Bahcall, and R. Davis, Jr., 'Solar Neutrinos: A Scientific Puzzle', *Science* CXCI/4224, pp. 264–267.

49. The tendency to revitalize religion is an issue we have already touched on in Note 31. Control, scientific management, flows of information and organisation, are topics which allow to reinstate concepts of providence, God the Father and faith. An example of the concurrence of interests in present science and institutionalized religion can be found in a conference held at the Bavarian Academy of Science in Munich, April 1978. Roman Catholic dignitaries and scientific dignitaries united here to make first steps towards an unholy alliance between ancient and modern religion. The discovery of control mechanisms in cells, organisms and molecules suggests to them the existence of an ordering power, system and/or centre in our universe. Guess who it is?

But similar tendencies are to be found elsewhere too. See e.g. R. Ruyer, *La Gnose de Princeton*, Paris, 1976; or F. Capra, *The Tao of Physics*, London 1975.

50. When I say 'system theorist' I mean here those people who make politics through model building. A nice selection of those mini-godfathers resides in Rome. It is, however, not the Holy See, as you might guess, but the Club of Rome.

51. T. Veblen, *The Theory of the Leisure Class*, N.Y., 1899.

   M. Weber, *Wirtschaft und Gesellschaft*, Part III, esp. Chap VI, J.C.B. Mohr, Tübingen, 1922.

52. J. Hirsch writes in his '*Wissenschaftlich-technischer Fortschritt und politisches System*':

   Dieser Wettbewerb wird dadurch verschärft, daß Produktions mittelhersteller mehr und mehr dazu übergehen müssen, ihren Kunden statt einzelner Maschinen und Geräte "Problemlösungen" und ganze Marketingsysteme anzubieten. Dabei kann die Offerte zum Hauptbestandteil der Lieferung werden, während das "harte" Gerät nur noch als ergänzendes Anhängsel erscheint.

   Suhrkamp Taschenbuch 437, Frankfurt a.M., 1971, p. 94. I translate:

   This competition is aggravated through the fact, that producers of means of production are increasingly forced to offer to their clients entire 'problem-solutions' and whole marketing-systems instead of single machines and equipment. It can happen that the offer becomes the main part of the bargain, while 'hard' equipment appears as a supplementary appendix only.

53. On the current impact of computers on science see e.g. D. Vahl, 'Chemistry without Chemicals', *New Scientist*, April 30th, 1970.

   In Germany the complete works of E. Kant were stored in a computer, the so-called KANT-computer, in the U.S. one hears of an enterprise where all known ancient Greek writings are to be stored by a machine.

   I start to wonder when the point will be reached where 'real data' and 'real

writings' become obsolete and all reference is made to the contents of an omniscient worldbrain.
54. Distribution of fiefs, beneficies and booty was at the core of medieval politics and consequently the main concern of ruling class' interest and science. In a comparable situation are the managers, whose central concern is distributive and not productive, as any organisational task is fundamentally a distributive task of duties and of gratifications. From this situation results a common feature between managerial and medieval interests and science, which will shape future science intrinsically.
   On the recurrence of some new version of the Middle Ages see also J.Y. Lettvin, 'The Second Dark Age'. In R. Clarke (Ed.), *Notes for the Future, op. cit.*, pp. 141–150; or L.S. Stavrianos, *The Promise of the Coming Dark Age*, W.H. Freeman Comp., San Francisco, 1976.
55. This tendency exists already today. An indicator for it is the pronounced interest in teaching material, i.e. one has to put one's ideas and research results in the form of a textbook to increase the chances of getting into the market.
   'Put all-space in a not-shall'', says James Joyce.
56. A step in this direction can be found in E. Leinfellner and W. Leinfellner, *Ontologie, Systemtheorie and Semantik*, Duncker and Humbolt, Berlin, 1976.
   See also J. Hirsch, *Wissenschaftlich-technischer Fortschritt, loc. cit.*, on the significance of management in science.
57. J. Bernal's book *'The World, the Flesh, and the Devil'*, *loc. cit.*, is indicative for this feature. Science, be it medieval, present or future science is monomaniac in its methods, answers and basic questions. To consider the world, the flesh and so-called irrationality as the three main enemies of man (homo?) is the product of such monomania. The world nurtured humans for millions of years well enough to keep humanity alive. The flesh is man and woman, as they are not brain exclusively. Stupidity and passions are not to be conquered by science, as the monomaniac suggestions of Bernal prove. Requiring the defeat of the world, the flesh and the devil has the same standing as requiring redemption from sin in another worldly paradise. The promise of paradise of any kind is a horror as it belittles present insufficiencies.
58. See S. Watanabe's 'Theorem of the Ugly Duckling', in S. Watanabe, *Knowing and Guessing*, New York, 1969.

RESISTANCE TO THE MACHINE

GEOFFREY PEARSON
University of Bradford

> The scenes of Manchester I sing,
> Where the arts and sciences are flourishing;
> Where smoke from factory chimneys bring
> The air so black, so thick and nourishing.
> Where factories that by steam are gated,
> And children work half suffocated,
> It makes me mad to hear folk, really,
> Cry, "Manchester's improving daily".
> Popular song, nineteenth century.

The theme of this essay is that of popular resistance to the machine and to the factory system of labour in the late eighteenth and early nineteenth centuries in England. I am, therefore, concerned largely with acts of violence which are usually thought of as 'senseless' and 'irrational' crimes — hooliganism and vandalism, smashing and wrecking — crimes which are thought to be pointless, all the more because they do not even submit to the guiding acquisitive principle of theft. I shall attempt to set machine-smashing and machine-breaking in the context of the motives and meanings which guided people in their attacks on machines and factories during the course of the 'industrial revolution': in other words, I shall assume that there is, in fact, sense and rationality in this behaviour which has been traditionally shrugged off as senseless, meaningless, motiveless, wanton, random, irrational and pointless. In order to show up the human intelligibility of resistance to the machine, I shall also discuss some earlier forms of 'resistance from below' — such as the food riot, and riot against the enclosure of the common lands — and I shall argue that machine-wrecking should be understood as a continuation of these earlier traditions of 'pre-industrial' resistance, rather than as an entirely new form of resistance to the newly emerging social order of the machine and the factory system of labour. Finally, I shall draw some points of connection between this early resistance to machines, and forms of struggle which are emerging against technology in the contemporary industrial world.

Throughout this essay I shall be drawing on a body of historical research which has re-drawn, even transformed, our understanding of the history of the common people of England — outstanding in this tradition are the works of Christopher Hill, Eric Hobsbawm, Edward Thompson and Raymond Williams (1). These writers share a common distinction which is distinctly socialist, and one of their principal accomplishments is that they drive for, and unearth, the experiences of the agrarian and industrial revolutions associated with the transition to capitalism as they were lived and felt 'from below' — that is, in terms of the material experience of the common people, rather than 'from above' in the committee chambers of high office.

The distinction between a history 'from above' and 'from below' is a crucial one. Riot against the machine, and against the rational 'improvement' of the land, is dismissed in a history 'from above' as mere 'social background' to the allegedly 'real' historical events which are described as the decisions and opinions of the mighty (judges, manufacturers, landowners, rulers, etc.) Not uncommonly, it is put down to the marginal activity of a few head-cases and lunatic hot-heads with criminal predispositions. In a revitalised history 'from below', however, the question of 'crime' and its definition emerges as something central to our understanding of the historical transition into capitalism.

These perceptual antagonisms between histories 'from above' and 'from below' are particularly sharp in relation to machine-smashing. Indeed, because resistance to technology appeared to evaporate, certainly after the mid-nineteenth century, then it is convenient to think of it as merely one of the 'teething troubles' of the immediate transition to industrial capitalism, a further reason to judge machine-wrecking to be something marginal. Against this, my argument will be that machine-wrecking was central to the transformation (as understood 'from below') and that the problems of environmental pollution and the destruction of traditional ways of life by the advance of machine technology (and its associated institutional forms) is endemic to a social formation which allows the technological form of production to be dictated by the principles of hard cash, rather than the historical needs of people.

Machine-smashing and the Industrial Revolution

Come all you cotton weavers, your looms you may pull down
And get employed in factories, in country or in town
For your cotton masters have found out a wonderful scheme
These calico goods they weave by hand, they're going to weave by steam.

So, come all you cotton weavers, you must rise up very soon
For you must work in factories from morning until noon
Your mustn't walk in your gardens for two or three hours a day
For you must stand at their command, and keep your shuttles in play.

(Nineteenth-century weavers' ballad by John Grimshaw of Gorton)

In the winter of 1842, when young Friedrich Engels came to live in Manchester — hot volcano city of the industrial revolution — he stepped into a world which had already been transformed by the machine. The cotton industry was the spearhead of the English industrial revolution, and machinery had been increasingly employed there since the late eighteenth century. The introduction of the power-driven loom, which accelerated from the 1820s, had virtually displaced the handloom weaver. The factory system of labour which accompanied the mechanisation of the textile industry had come to rule over and rule out the small-scale domestic economy which had previously been the dominant form of production in the small towns and valleys of Lancashire. Lives were transformed by these 'dark satanic' developments, and the extent to which the factory system destroyed traditional patterns and rhythms of work is described in numerous songs and ballads of the period, such as John Grimshaw's lament. Although Grimshaw's song misrepresents the scale of the changes: people did not work in factories 'from morning until noon'; they were tied to the machine for as much as sixteen hours a day. We can only assume that Grimshaw was struggling for a rhyme.

Manchester was the symbol of the new age. Almost everyone who was anyone, it seems, came to Manchester in the mid-nineteenth century to marvel at the wonders of the new industrial system — Dickens, Disraeli, de Tocqueville, Carlyle (2). But they usually went home shocked and horrified by the brutality of Manchester. Unless, that is, they were particularly thick-skinned, or blinded by the glitter of wealth which could pour out of the thundering factories. "Manchester is the chimney of the world" is how Major-General Sir Charles James Napier, appointed to command the northern district at a time of massive political disturbance in the manufacturing districts, put the matter. But who, he added, wants to live in a chimney? He might also have mentioned (others did) the stinking, polluted rivers, the excrement piled high in the streets, the squalid cellar homes, or the hordes of unemployed men standing in the streets which greeted Engels' arrival during the depression of 1842.

What is truly extraordinary is that the changes which produced this monstrosity — in Ancoats, a working class district of Manchester, as few as 35 children out of 100 born would survive until their fifth year — can be described by historians with such calm rhetoric. In a 'history from above' the creation of the factory system is described as the steady erosion of inefficiency by the captains of industry and their new rational methods. It is a history of progress and buoyancy. It tells of the enterprising men — Arkwright, Crompton, Peel, Hargreaves — whose inventions, their mules, carding frames, spinning jennies, etc., and whose capital was to transform the world.

What a history from above is more shy about is that the common people frequently attacked both the machines and their inventors, tore down and burned their mills, and drove these 'great heroes' of the industrial revolution out of the locality. If a history from above does mention this opposition at all, it puts it down to 'ignorance'. The story of this bitter resistance, the sense and meaning of riot and vandalism, is reserved for a history from below. A history, that is, which drives for the viewpoint of the common people and how the machine contributed to the destruction of their way of life.

Attacks on machines took place from the 1760s, when Hargreaves' spinning jenny was repeatedly smashed, and mills which used the jenny were turned over. According to traditional accounts, Hargreaves was chased out of the neighbourhood and his promotor, the factory owner 'Parsley' Peel, retreated from North-East Lancashire in disgust, taking his capital to another area where he hoped the work force would be more sensible. Peel's mills near Blackburn and Accrington were completely destroyed, one having been already rebuilt after attacks by machine-smashers only a few years before.

There were many more attacks on machines. One of the most intense periods of disturbance was during the War in Europe, when from 1811 to 1813 various kinds of new machines were attacked in Nottinghamshire, Yorkshire and Lancashire. The attackers, known as the 'Luddites', claimed leadership from a probably fictitious 'General Ludd'. To mention one more instance, there was a great uprising against the machine in North-East Lancashire in 1826 when power-looms were destroyed in the cotton towns. Machine-smashing was not restricted to the textile industry, of course. In 1830, there was an explosion of popular discontent throughout southern and eastern England, and other parts of the countryside, when the rural poor attacked the hated threshing machines and set fire to hay ricks. Known as the 'Swing Riots', these rioters attacked the machines in the name of 'Captain Swing' (3).

Considerable dispute surrounds the question of how to interpret the activities of the machine breakers. One dominant traditional form of argument describes these outbreaks of violence as the sporadic, unpredictable and senseless antics of the riff-raff — the mob, *'la canaille'*. It asks, "Why does the mob riot?" And it answers: "For the fun of it," or simply "Who knows?" There is a powerful line of historical continuity between this response to 'the mob' and modern accounts of hooliganism and crime as senseless (4).

Another traditional line of reasoning suggests that loom-breaking, etc., was the desperate action of people pressured and frustrated by the strains of social change into wild and lawless behaviour. This historical form continues, however, to describe the mob as senseless. Its only concession is that machine-breakers are victims of forces outside their control which cause them to behave senselessly and randomly. For example, Smelser, in his Parsonian sociological analysis of the industrial revolution in Lancashire, points to shifts and alterations in the organisation of the economy, factory labour and family life which (he argues) released the bind of steady socialising influence and led to pathological and unsocialised behaviour, that is machine wrecking. He describes it as "a relaxation of the most basic controls over socialised behaviour" and as "violent and bizarre symptoms of disturbance" (5).

In both cases — whether it is the snobbish refusal to allow any meaning whatsoever to mob riot, or the paternalistic response of judging the behaviour as a pathology — the idea that 'the mob' represents purposive, rational conduct is banished, out of sight, out of mind. However, these dismissive accounts inevitably fall prey to a contradiction in that, whilst their controlling argument is that machine breaking is irrational and pointless, it is exceedingly difficult to portray the factory system in anything other than a dark light. And it is precisely this dark aspect of the factory and its machines which, in many important respects, guided machine-breaking and gave it its rationality.

Take, as an illustration of this contradiction, some statements by the distinguished historian of the eighteenth century, H. J. Plumb. His account of machine-smashing is, admittedly, schematic, since in the tradition of 'history from above' Plumb is always restless when describing social conditions, anxious to get on with the 'real business' of intrigue between rulers. Nevertheless, his account is plagued with contradiction. At one point he states, for example, how the new machines in textiles "revolutionised the production of yarn and brought to the weaver an age of golden prosperity which was to last for a quarter of a century" (6). The statement is not so ridiculously wrong-headed as it might at first appear. The earliest phase of mechanisation involved the introduction of improved *spinning* machinery which greatly increased the output of yarn and hence the demand for weavers – and the weaving looms were, as yet, unmechanised. The complexities of this early period are best set out in Thompson (7) where he shows that, although the idea of a 'golden age' is a myth which overstates the prosperity of the weavers in this period (roughly from the 1780s until the early 1800s) there was, nevertheless, a period of boom in handloom weaving. But from then on, the condition of the weavers was steadily attacked, so that looking back the earlier period might appear as a 'golden age'. As Thompson expresses the matter: "The history of the weavers in the nineteenth century is haunted by the legend of better days" (8). But even in this earlier period of relative prosperity, cotton mills and machines had been attacked and there is nothing to justify Plumb's rosy optimism of an overall improvement in living standards.

The contradictions deepen in Plumb's account as the antagonistic encounter with the factory system progresses. The worsening conditions are inescapable – the increasing exploitation of children in the mines and factories, the erosion of the more intimate and flexible forms of production which preceded the factory, the repression of working class movements and early trade union organisation, the contracts (occasionally as long as seven years) which bound men to their factory masters under risk of imprisonment, the massive increase in the number of capital statutes which could send men to the gallows for a whole number of newly defined crimes, the destruction of the makeshift economy of the poor which followed from the enclosure of the common land which robbed the common people of their common rights — 'rights' which were often re-defined as 'crimes' (poaching, trespass, wood theft). Plumb mentions most, if not all, of this. He also writes that workmen "viewed with deep suspicion the barrack-like factories whose long and regular hours savoured to them of the prison" (9). Under the pre-industrial systems of production, men had been much more in control of the rhythms of their lives. One should not romanticise this state of affairs — as if every man had been an independent, unalienated and free man — but one should recognise, nevertheless, what was involved in the changes in production systems. Before the factory, workers had customarily observed 'Saint Monday' as a day of rest, often taking Tuesday as a holiday as well, then working flat out to finish off the week's work. Factory discipline and factory time changed all that. As Plumb describes the factory:

The hours of work were fourteen, fifteen, or even sixteen a day, six days a week throughout the year, except for Christmas Day and Good Friday. That was the ideal time-table of the industrialists. It was rarely achieved, for the human animal broke down under the burden; and he squandered his time in palliatives — drink, lechery, blood-sports. Or he revolted, burned down the factory, or broke up the machinery, in a pointless, frenzied, industrial *jacquerie*. (10)

What is extraordinary is that Plumb, having set out the grounds for popular discontent as explicitly as this, continues to describe machine riot as 'pointless'. The same contradiction is found in other writing on this period. In Bythell's work on the handloom weavers, he writes about the 1826 power-loom smashings: "Their Luddism of 1826 was a blind display of hatred against an improved machine which must be destroyed before it took away the old weaver's livelihood", and he refers earlier to these same riots as an "act of blind vandalism" (11). But, we must ask, what is blind about an action which strikes out to destroy something which threatens your own destruction?

What is expressed in these simple contradictions is that it is somehow intolerable to allow rationality to popular resistance and popular violence. It is much more comfortable to write off the machine breakers as a tiny pathological fringe of hot-heads, and to write a cosy history in which the mass of people are said to quietly bear the yoke of the new factory system. That is, as if people understood the superior rationality of the machine, and accepted the injustices which flowed from this superior rationality for the sake of 'progress' and 'improvement'. It is in these ways that historical thought has simply followed the contours of power. Although, as we have seen, even in its language it continues to express — against the will of the author as it were — the bitter contradictions and struggles which constituted the real history of the period.

What is, perhaps, an even more fundamental difficulty with these accounts is that we know that within their own community the machine wreckers often had a heroic status. In striking out at the machine, they were striking out against their oppression within the developing system of factory labour — the relentless monotony of the pace of the machine, the relentless gaze of the factory master and overseer, the relentless tick of the factory clock.

Above all, what must be remembered is that just as the human significance of time (12) was changed by the new rhythms of the factory and the machine, so were many other aspects of culture. In one sense, the arrival of the machine and the factory was the first direct encounter by working people with the *material experience* of hard-headed utilitarian philosophy and political economy. The dominating principle of hard cash, of buying cheap and selling dear, also turned the human relations of the factory into commodity relations and transformed men into objects or cattle. So that when Plumb writes about the "human animal" breaking down under factory conditions, it is necessary to remember that it was precisely the logic of the new systems which reduced men from human to animal status. Language both mirrored and informed these transformations: so that the men and women who worked the machines came to be called, simply, 'Hands'. The word 'common' also underwent a transformation, away from the sense of things 'held in common' or a 'commonwealth', towards a revaluation of certain (particularly plebian) practices and customs as 'common', meaning 'vulgar' (13).

The experience of these transformations should not be underestimated. Resistance to the factory and its logic continued well up into the 1840s, and the Chartist movement in some of the Pennine cotton towns at this time carried as one of its demands that each man should receive, as a birthright, a small-holding of land — *and* they began staking out Pendle Hill for just this purpose!

The logic of the factory and of enclosure worked in an opposite direction. The independence of working men who had access to small patches of land was, in utilitarian terms, understood as dangerous and wasteful. As late as the

Select Committee on Commons Enclosure of 1844 we find the assertion that the enclosure and "improvement" of the land would also bring an "improvement" in the character of the people; and that "the unenclosed commons are invariably nurseries for petty crime." (14) What is meant by this is that if they had access to land, men would tend their pigs, or their cow, instead of turning into the factory. In the Poor Law Reports of 1834, the culmination of utilitarian principle, there are attempts to calculate how big (or, rather, small) an allotment of land should be given to working men. It was agreed that a small allotment for vegetables provided a working man with an innocent and sober amusement, and also enabled him to feed his family, thus keeping down the likelihood of wage demands. But the utilitarian logic also dictated that the Poor Law Commissioners should try to calculate what size of an allotment (that is, garden) would provide amusement and some food without causing the working man to keep away from work, or without causing him to waste too much of his valuable body energy, his labour power, which was the rightful property of the factory owner — and not the man himself. There was a recognition in these Poor Law debates that food (largely potatoes) equals energy, equals improved labour power. But that digging a potato patch expends energy, and that it is more enjoyable to spend energy digging the earth to provide food for one's family than to rise at the crack of dawn and watch over a noisy and dangerous machine for someone else's benefit (15). It is a significant moment of illumination of the nature of the contest which was involved in the advancement of the machine and factory.

It is thus in the nature of this historical transition — which involved profound changes in human values as well as a technological break — that it is utterly impossible to go along with Smelser, for example, and describe the machine breakers as disorganised and unsocialised beings. Instead, we need to ask: "Socialised according to what standards?" And in particular: "Socialised according to *whose* standards?" The machine smashers were not only courageous and determined in their attacks on the machines, they were also sometimes very discriminating. In early attacks on the spinning jenny during the 1760s, for example, the rioters sometimes spared small machines which had less than twenty spindles — that is, machines which could be operated in small cottages on the established principle of the domestic economy, and which were for this reason regarded as 'fair' machines. And that is hardly behaviour which could be described as 'bizarre' or 'frenzied' or 'blind' or 'irrational' or 'unsocialised' or 'random'.

The great power-loom smashing of 1826 provides other examples of these kinds of rationality (16). The outbreak was a particularly determined one which was confined to the cotton towns around Accrington, Blackburn and the Rossendale Valley and which was eventually put down by violent Army reprisals. It is even possible that we should think of it as a general insurrection against the machines in these towns (17). Be that as it may, the rioters had considerable sympathy within the community. Robert Peel, eventually to become Prime Minister, a man whose fortunes were founded on cotton, found it necessary to rebuke local magistrates for not taking sharp action against the rioters — whether he was motivated as a politician, or as a man with property stakes in the towns, is not clear. The reluctance of magistrates to act decisively was not uncommon, reflecting the fact that local opposition to machinery was so widespread. For example, small masters feared and hated the new machine as much as working people. Without sufficient capital to invest in machinery, the machines threatened to kill their small businesses — a fact which could provide some unexpected alliance of sympathy between the poor and the lower echelons of the owing class. Even before the 1826 attacks on the power-looms had started, rioters had been stopped by troops — the most solid embodiment of State power. But when the troops heard the complaints of the weavers, instead of opening fire they opened their knapsacks and gave their food to the rioters. The troops then moved on, the weavers held a meeting amongst themselves to decide whether or not to continue: "Were the power-looms to be broken or not?" "Yes", it was decided, "they must be broken at all costs" (18). When the first mill was attacked — Sykes Mill at Accrington — the first thing to go, smashed by a woman — was the tyrannical factory clock, the hated symbol of the slave rhythms of factory labour (19). By nightfall, it is said that there was not a power-loom left standing within six miles of Blackburn. The riots spread to other nearby towns and within three days more than a thousand looms had been destroyed.

There is one sense, of course, in which the loom-breakers had maybe got it quite wrong. 1826 was a year of slump, and the grievances of the handloom weavers, it can be argued, derived from the trade cycle and not from the machine. Nevertheless, although outbreaks of machine smashing coincided frequently enough with periods of trade recession, the argument is finally unconvincing, one which lends itself too easily to a crude economic determinism which ignores the cultural developments which are decisive for an understanding of machine riot. (See Note 44.)

Nevertheless, it is important to recognise that different outbreaks of machine-breaking could, and did, have different motives. Quite often, for example, what was being contested was not the principle of the machine itself, but the level of wages, etc. Attacks on machines provided a way of getting at an employer so as to encourage him to raise wages, improve working conditions, take on more men, or whatever. It is this kind of motive for machine-breaking which Eric Hobsbawm has described as "collective bargaining by riot", a primitive form of trade union struggle.

These "primitive rebels", to use another of Hobsbawm's suggestive phrases, are said to be primitive in the sense that they had not yet found a mature political vocabulary, nor a mature political strategy, within which to phrase their discontent and as a means to fight their struggle. This line of thinking urges us to conceive of the Luddites and other machine breakers, then, as precursors of the mature forms of labour organisations.

Hobsbawm's distinctions are important if only because they caution us not to oversimplify machine-smashing, and not to think of it all as direct resistance to machinery as such. But it is nevertheless a form of historical reasoning which can, in its own way, belittle, discredit and misconstrue the motives and actions of the machine-smashers. In claiming machine breakers for the history of the labour movement, for example, it causes us to compare their frail outbursts with the more 'mature' forms of struggle — trade union organisation, collective bargaining, the strike, the general strike — which only serve to underline the 'primitivity' of the machine smashers. As working class consciousness emerges, it tells us, then men set to one side these infantile forms of resistance to exploitation and "see the light".

It is thus an inherent danger of the concept of "primitive rebellion" that it can lead us to think that the intelligibility of the machine breakers' struggle only emerges as history unfolds; that its intelligibility is derived only from the subsequent history of the labour movement. As if the intelligibility of the machine-breakers emerges only when they are already dead. And when they are dead, then we can come to see them in their 'true' historical significance, the vanguard of an emerging proletariat.

But the heart of the matter is that the machine-smashers were not a proletariat. On the contrary, they would as often as not be people from the countryside, with its own distinct inherited traditions. People who looked at the factories — shaking and trembling with their violent energy — and who did not like what they saw. It is an important truth that there *were* some people around at the time of the Luddite disturbances from 1811 to 1813 who were part of the general mood of insurrection which surrounded the period and who expressed more forward-looking political opinions. People who saw in the machine the possibility of liberating men and women from the burden of physical labour, and who could be thought of as early forerunners of a later working class consciousness (20). But they were the exception rather than the rule. Thompson, with his usual lucidity, expresses the matter thus: "For those who live through it, history is neither 'early' nor 'late'. 'Forerunners' are also inheritors of another past. Men must be judged in their own context" (21). And for this reason, if we are to set machine-breaking in its full context, it is necessary to turn back briefly to some of the inherited traditions of the machine-breakers' world, a world which was entering a rapid eclipse.

The Traditions of the 'Crowd' and 'Riot'

<table>
<tr><td>
Farmers taek
nodist form
This time be
fore It is to
let

Be fore
Christ mus
Day sum of
you will be
as Poore as
we if you
Will not seek
Cheper
</td><td>
This is to let you no We
have stoel a Sheep, For which
the reason Was be Cass you
sold your Whet so dear and if
you Will not loer prices of
your Whet we will Com by night
and set fiear to your
Barns and Reecks gentleman
Farm mers we be in Arnest
now and That you will find
to your sorrow soon.
</td></tr>
</table>

(Anonymous letter fixed to the pillory, Salisbury market 1767.) (22)

Luddism and the other machine-breaking riots were not the first burst of a new class anger from an infant proletariat. Rather than being based on the new, emerging traditions of the new-fangled men of that new class, they were more likely to be based on older traditions. Traditions, that is, which looked back to the domestic economy, to the common lands, and to former traditions of radicalism.

'Riot' and 'crime' had been an important feature of the social life of 18th-century England. The 'crowd', or the 'mob' could take as its object the defence of any number of rights and customs — the fair price of food, fair access to the common lands, or in defence of smugglers and smuggling which was so common in England, as in most parts of Europe, to be thought of as a 'right' rather than a 'crime'.

In fact, where one drew the line in 18th-century England between what was a 'crime' and what was a 'customary right' depended very much on one's position in the social order. For example, an activity called 'wrecking' was a common 'crime' in those coastal regions where ships most frequently ran aground and were wrecked (23). 'Wrecking' was a form of coastal plunder in which local people would salvage what they could of wrecked ships and their cargoes for their own use, and it formed an important part of the local economy in some coastal regions. So that, from the point of view of the local people, 'wrecking' was an important, even indispensable, part of local life. Whereas, to shipping merchants it was a criminal nuisance, and merchants petitioned the government throughout the 18th century to act more severely against wreckers. And as a result, an Act of 1753 made 'wrecking' a crime punishable by death — although that did not stop wrecking.

It is a similar story in relation to a large number of other customary activities of the poor and the common people — taking hares and rabbits from fields, taking gravel and turf; taking fish from streams; collecting wood in the forests. The notorious Black Act of 1723 was one bloodstained piece of legislation, drawn up with extraordinary speed and carelessness by Walpole and his ruling clique, to contain these kinds of offences. The precipitating factors were complex, as Thompson shows in *Whigs and Hunters*, his masterly analysis of the Black Act, involving amongst other things attempts to curry favour with the new Hanoverian monarch whose deer parks at Windsor were in a state of collapse (24). Local people in that region had begun to reassert their rights in the forest, as opposed to the rights of the deer. Thompson suggests that people may have been looking back to the days of the

Commonwealth when "the deer had been slaughtered wholesale, the Great Park turned over to farms, and the foresters had enlarged their 'rights' beyond previous imagining" (25). And although there was nothing like an insurrection on that scale in the 1720s, the struggle between men and deer had been re-engaged — to the disadvantage of the deer. These conflicts should not suggest simply strong feeling against the monarchy. Deer were a damned pest which ate and trampled on crops. Sometimes it is necessary to resurrect very specific details of the local economy in order to understand these struggles. Around Farnham, for example, timber rights were crucial. Farnham was a hop-growing area, depending on a supply of good, strong poles to support the hops: "but if deer cropped ash saplings they grew up bent and unusable for poles" (26). Men and deer were enemies: unless, that is, the men were privileged men with the right to hunt deer and eat venison.

The Black Act, which was a response to these struggles, made at least 50 distinct offences punishable by death: principally, going about the countryside armed and in disguise with one's face 'blacked' (hence the 'Black Act', although the Act was obviously black in other respects too); hunting, wounding or stealing deer; poaching hares, rabbits or fish; damaging fishponds; cutting down or damaging trees; maiming cattle; sending anonymous threatening letters; setting fire to houses, barns, etc. Although it was unprecedented in its venom, the Black Act was just one of many similar enactments in this period: during the 18th century there was an astronomical increase in the number of offences punishable by death (27). So that, whereas a 'history from above' might describe the development of Law in 18th-century England as the rationalisation of property relations, providing a coherent system for the regulation of trade and commerce; 'from below' the Law was experienced as a lash of the whip, the threat of transportation, the gallows at Tyburn, and the awful sight of the bodies of convicts swinging in chains. These were the delicate means by which the gentle rulers of Merrie England set out to terrorise the population into conformity with the steady encroachment of the new commercial values and the new property rights.

What precisely did these encroachments involve? Take first the example of land: under a pre-capitalist arrangement, land was frequently a place where a number of coincident use-rights intersected. This man would have the right to graze cattle here; that man would have the right to take wood from there; this one would have the right to gravel or turf; that one to hares; another would have the right to timber; etc. Sometimes these rights were defined by copyhold — that is a legal arrangement which might have been established many years ago. More commonly perhaps, they were simply honoured as customs which went back, as the people would say, 'time out of mind'. But this messy arrangement of coincident use-rights was anathema to the emerging forms of capitalist ownership through which land was transformed into a commodity:

> During the eighteenth century one legal decision after another signallled that lawyers had become converted to the notions of absolute property ownership, and that . . . the law abhorred the messy complexities of coincident use-right. And capitalist modes transmuted offices, rights and perquisites into round monetary sums, which could be bought and sold like any other property.(28)

This was also a period in which the great country houses and country seats of England were being built — and Thompson jokes that it was not much fun for those being "sat upon". The volcanic wealth of the new commercial interests — based on stock jobbing, or on the slave trade for example — was unimaginable in a traditional agricultural economy. And as these new city spivs exported their city wealth into the countryside, building luxurious mansions and parks, local farmers and yeomen were squeezed out. William Cobbett writes in his *Rural Rides*, for example, about Lord Aylesbury's parks: he "seems to have tacked park on to park, like so many outworks of a fortified city. I suppose here are 50 to 100 farms of former days swallowed up" (29). The case was in no sense exceptional. Wealth was beginning to accumulate into lumps, and the land with it. Soon the complaint would be heard, and Cobbett would be one to voice it, that the people were also accumulating into 'lumps' or 'heaps' — that is, in the factories and cities.

It is a vital historical transition surrounded by conflict amounting almost to guerrilla warfare. But often, at the centre of this warfare, it was not the very poor, and the poor may have had little to do with the struggles which surrounded the Black Act, for example. It was the middlemen — the small farmers and yeomen farmers — who often had most to lose, who were at the centre of the disturbances which surround the Black Act, and who were the 'Blacks' who visited the houses and parks of the new rich, who ripped down their fences and attacked their deer:

> We appear to glimpse a declining gentry and yeoman class confronted by incomers with

great command of money and of influence, and with a ruthlessness in the use of both. . . Their families must have had a rich and tenacious tradition of memories as to rights and customs. . . But these farmers had no money from sinecures or killings on the stocks with which to manure their lands, and they remained stationary or declining, with a traditional economy, while the new rich moved in all around (30).

But although, in this instance, the poor may have been marginal to the disturbances (and even so they were drawn into the conflicts, and labourers figure strongly in the prosecutions brought under the Black Act) there were other areas of resistance where the poor occupied the hot centre. One such form of resistance was the food riot, a persistent interruption in the history of 18th-century England. Thompson has shown how the 18th-century food riot was no random, hit-or-miss affair, but an activity tuned to a precise 'moral economy' (31). Pre-capitalist relations between producers and consumers on the local markets had been hedged about with many regulations and customs which safeguarded tangible rights of the common people and the poor. For example, corn could not be sold standing in the field to a large buyer; nor was it proper form for farmers to withhold their corn in the hope of rising prices. Corn had to be transported to the local market, and there large buyers were required to delay their deals until common people had made their small purchases. Market bells signalled the restricted period in which the poor could buy first, and weights and measures were also supervised. Most of this regulation of the market was based on custom or common law, although it was sometimes regulated by Statute. For example, "an enactment of Charles II had even given the poor the right to *shake* the measure, so valuable was the poor man's corn that a looseness in the measure might make the difference to him of a day without a loaf" (32).

These regulations were, naturally, constantly evaded and ignored: it is the 'natural' privilege of ruling groups to break their own rules without these infractions being called 'crimes' and punished with severity. Although, to press the point further, these were not 'their own' rules: they were the rules of an older, dying pre-capitalist formation which the new men, giddy with the prospect of wealth which the new capitalist values promised, had no interest in enforcing. The temptations to break the market customs were many: particularly the growing demand from the cities and large towns, in particular hungry London. Selling in bulk to a large dealer, selling corn as it stood in the field, or hoarding corn in anticipation of a better price — these were also more 'rational' market principles which maximised the possibilities of gain for the farmer's labour and investment.

However, another 'rational' and 'natural' consequence of these rational and natural developments was that the poor did not see eye to eye with the farmers and millers. In times of scarcity, or when prices rose dramatically, or when poor people without sufficient food watched as large convoys of food trundled along the roads towards the growing cities, 'riot' was a common means of restoring the old regulations. These riots were sometimes, although not always, highly disciplined affairs in which the people attempted to remind businessmen of the old customs and encouraged farmers to sell at a fair price. Commonly enough, the 'crowd' engaged in direct action on these occasions taking the corn from the farmers by force and distributing it amongst themselves at a fair price (known as 'setting the price', similar to the French '*taxation populaire*') whereupon the money, and in some instances even the empty sacks, were handed over to the farmer as his rightful due.

At other times, food riots were conducted in an atmosphere of carnival: farmers or millers were ragged to the accompaniment of 'rough music', shops or stalls were ransacked or burnt, and in the pandemonium of the Great Cheese Riot at Nottingham Goose Fair in 1764 whole cheeses were rolled down the street.

One of the more interesting features of this festive version of riot was the way in which ancient stocks of offensive ritual — principally the ceremonies of 'misrule' employed at the festivals and holidays of apprentices and students in order to mock their elders, together with the 'rough music' which had been used traditionally to register public disapproval of lechery and other forms of sexual impropriety — were turned to political ends in the 18th-century food riot (33). In these ancient rituals of disapproval, various kinds of ceremonial mockery had been employed — particularly by the youth of the community — against a number of traditional targets such as older widowed men who took a young girl for a second wife, girls who were known to be promiscuous, nagging wives, hen-pecked husbands, and wife-beaters. Terrifying or obscene effigies were paraded in the streets, an effigy of the victim would be dragged seated backwards on a donkey, there was pantomime, faces were blacked and disguised, fancy dress was worn, there were stink-bombs and horse-play — and invariably all this would be accompanied by 'rough music' made with pots and pans, tambourines, bells, whistles and horns, rattles and jew's harps.

Thomas Hardy provided a brief account of this kind of riotous ceremony, known as the 'skimmington ride' or 'riding the Stang', expressing public disapproval of the perceived misconduct of a local couple in *The Mayor of Casterbridge*. The Wessex 'skimmity-ride' involved tying effigies of the couple to a horse or mule, and rough music was an indispensable part of the mockery: "the din of cleavers, tongs, tambourines, kits, crouds, humstrums, serpents, rams' horns, and other historical kinds of music" (34). To give one final example of this stock of profane ritual, whereas girls of marriageable age would have their houses decked with the 'May Bush', young ladies who were thought to be too promiscuous might be visited in the night and find outside their door the gorse bush – sometimes known as the 'smelly bush', which makes the meaning of the ritual clear enough. These pre-industrial customs, which had traditionally provided an arena for youthful merry-making and 'piss-taking', were characteristically vulgar, although this was part of their profane force.

It is, then, an interesting and significant shift which takes place in the 18th-century English food riot, when such rituals as these were appropriated to a more generalised struggles of an economic and political character. Further evidence of the continuing, but changing traditions of protest can be seen in the Rebecca Riots in West Wales between 1839 and 1844. In Wales, the enclosure movements came later, and local grievances about tithes, grazing rights, timber cutting, and more general political agitation around the hated Poor Law and the demands of Chartism, reached a flash-point on the question of toll-gates. These events saw the reappearance of a figure not unlike 'Mother Folly' – one of the familiar characters amongst the ancient 'Lords of Misrule' – when men dressed as women, with their faces blacked or disguised and calling themselves 'Rebecca and her Children', attacked toll-gates, fences, salmon weirs, workhouses and other material symbols of the new order. These widespread riots often resembled the ancient custom of '*ceffyl pren*', the Welsh equivalent of 'skimmington', with its usual accompaniments of 'rough music' and pantomime (35). And although the *ceffyl pren* was turned to political ends, Rebecca still found time to visit the old haunts, to intervene in family feuds, to return illegitimate children to their rightful parents, or settle accounts with wife-beaters, providing an important indication of how deeply the customs and traditions held in the minds of the common people, and also of the extent to which these forms of riot continued the traditional functions of social disapproval and community control which had been exercised through rough music and misrule. As John Gillis describes one aspect of the transition: "In their desperate defence of the just price, eighteenth-century English crowds often transformed rough music, traditionally expressive of moral indignation against lechers, into instruments of class conflict. The miller's legendary prowess with young women who came to his mill became a convenient metaphor for a different kind of exploitation, economic rather than sexual" (36).

'Riot', then both in its disciplined and carnival forms, was a well established theme of resistance in the eighteenth century. As often as not, it was not necessary to go the whole hog, and simply the threat of riot would be enough to remind magistrates of their obligations to enforce the old customs, or to intimidate traders into selling at a 'fair price'. Preachers might urge the poor that prayer was the most effective means of surviving food shortages or price rises, but as Thompson remarks: "The nature of gentlefolks being what it is, a thundering good riot in the next parish was more likely to oil the wheels of charity than the sight of Jack Anvil on his knees in church" (37).

Riot or the threat of riot thus formed part of the irregular democracy of the countryside. In his essay 'The Crime of Anonymity', Edward Thompson shows how anonymous threatening letters provided one widely used means of dialogue between the 'crowd' and the millers, farmers and magistrates (38). Peter Linebaugh has shown, in another context, how often a poor man sent to the gallows by a man of means should threaten to haunt his persecutor (39). The threat seems laughable by modern standards, but in worked often enough to make sense, and the rich man would call off (or buy off) the long arm of the law in order to avoid being troubled by bad spirits, or a bad conscience. The prospect of public ridicule, as we have seen in the brief excursion into the traditions of rough music and misrule, provided another means of putting pressure on powerful members of the community. At other times the threats were more blood-curdling: "This will all come true . . . kill the over Seeer . . . Tom Nottage is a dam Rouge . . . kill him for one there is 4 more we will kill . . . sink your Flour to 2 and 6 a peck set fier tu it and burn it down . . . burn up all the Mills . . . burn up all over thing an set fier tu the Gurnray" (40). The threats were nothing if not extravagant, and sometimes they even came in verse:

> This comes to let you know that there is a small Army of us
> Upwards of 3 thousand all ready to fight
> & I'll be damd if we don't make the Kings Army to shite
> If so be the King and Parliment don't order better
> We will turn England into a Litter
> & if so be as things don't get cheaper... (41)

And so on. This extravagance could get quite out of hand (during the weavers' struggles against the spinning jenny in Lancashire in the 1760s there were threats to "pull down" whole towns!) and it seems unlikely that many would believe such enormous threats. However, that was the *genre*: a boast that an army of men was sworn in and ready to fight and kill and maim, an army which would rather "die by the sword as by hunger" (42) was a warning shot from the crowd. These were poor means and inadequate means, by any standards, and yet they were so often the only means available by which to activate the shop-soiled traditions of paternalism. Thompson writes of threatening letters, and similar forms of protest, that they "may sometimes be seen as intrinsic to proto-democratic forms of organisation, deeply characteristic of eighteenth-century social and economic relations" (43) and we can see in the *genre* a continuity of struggle which joins the superficially different issues, for example, of food prices or opposition to machinery. These were defences of the weak which, at different times, could (and did) coerce local authorities into subsidising food prices on a local basis, which could provoke local magistrates to speak out against the uncontrolled advance of the machine, or which could pressure landowners to admit other common rights ... at least for a time.

However, as far as these forms of popular resistance were concerned, their time was up. So was the time of the forms of social organisation from which this resistance claimed its legitimacy. If we once more take the arrival of Engels in Manchester as a datemark, both were by then as good as dead. Population changes spell out the scale of the changes. In 1773, Manchester had been a town of 24,000 people. In 1801, its population had almost trebled to 70,000; by 1831, it had doubled again to reach 140,000 with almost a 45% increase in the decade from 1821 to 1831. And so it went on. In 1841, there were 217,000 people in Manchester; and by 1851 — the landmark year in which the census revealed that now the urban population of the entire nation exceeded the rural population — Manchester's population was a quarter of a million (44). If one takes the whole of the surrounding urban conurbation of Manchester, then the figure is close to half a million people living together in one large 'lump'.

But more had changed than just the scale of things. Most importantly there had been the emergence of Chartism demanding a People's Charter involving electoral reform based on universal manhood suffrage, which signalled a new kind of consciousness within the working class. The summer of 1842 had also seen a massive industrial and political upheaval throughout the cotton manufacturing districts of Lancashire, with 50,000 workers on strike in Manchester alone. These were the 'Plug Riots', so named because striking workers marched into factories and removed the plugs from their boilers and furnaces, thus stopping the factory and enforcing the strike. After considerable political disagreement within their ranks, Chartist leaders came in behind the strike which moved towards a critical political confrontation, demanding the Charter as the condition for returning to work.

In immediate terms, the Plug Riots strike was a total failure, involving arrests on a wide scale, and the establishment of permanent military detachments in the areas. So that when Engels stepped into this transformed world, he was stepping into a world where not only was it reported in some of the industrial towns that "numbers kept themselves alive by collecting nettles and boiling them down" (45) but also fifteen hundred working class militants were in prison. The people associated with Chartism and the Plug Riots were different kinds of people from those who had broken the machines, with different demands. For better or worse, many of them had come to accept the factory and the machine, and had resituated their politics accordingly. Although that is to state the transition too abruptly, for some of the old memories, traditions and passions were to take a long time dying, and as I described earlier, in the smaller outlying cotton towns some of the Chartists still harked back to the 'pre-industrial' demand that each man should have a smallholding of land, and in 1842 they were staking out the surrounding countryside in preparation for the millenium (46).

Also, on the question of machinery, there were splits within the Chartist ranks. For example, W. Cooke Taylor in his *Notes of a Tour in the Manufacturing Districts of Lancashire* (1842) found that at Burnley "the block-printers and handloom weavers united to their Chartism a hatred of machinery, which was far from being shared by the factory operatives" (47).

There was also a split on the question of the use of physical force, apparently reflecting the same divisions of interest between the older crafts and the new factory operatives, whereby those engaged in the older crafts "urged an immediate appeal to arms . . . and the expediency of burning down the mills, in order to compel the factory-hands to join in an insurrectionary movement" (48). At Colne, a mill had been burned down a few days before W. Cooke Taylor's arrival, and in neighbouring Burnley he found people expressing the hope that this would be imitated, "while the heaviest curses were bestowed on the factory-hands of Colne for having heartily exerted themselves to check the conflagration, and to supply water to the engines" (49). Meanwhile at Padiham, another neighbouring town, Cooke Taylor came across a sentiment which harked back to the rural 'Swing' riots of 1830: "I heard a man in the open streets go beyond even the violence of Burnley, and, amidst the cheers of some scores, express an eager hope that 'Captain Swing would take command of the manufacturing districts' " (50).

No, resistance to machinery was certainly not dead by 1842. In fact, perhaps we can more usefully say that the Plug Riots and the time of Engels arrival in Manchester stand as a bridge between two traditions: those of the dying crafts and attitudes of pre-industrial England, and the emerging de-skilled factory workers who formed the mass of the new 'proletariat'. The political heritage with which Engels came to be associated, of course, turned its back on the old traditions, and embraced the new. This is entirely understandable: to do otherwise at that historical conjuncture would in all probability have meant political suicide in the face of relentless and seemingly irreversible changes. Nevertheless, given the nature of what had been involved, we can perhaps only marvel at the remarkable insensitivity of that well-worn phrase from the *Communist Manifesto* which celebrates the new system as something which "has thus rescued a considerable part of the population from the idiocy of rural life" (51). That, to put it bluntly, is not how the machine-breakers would have judged the matter.

The Last Word

> . . . a place called Manchester, which has now disappeared
> (William Morris, *News from Nowhere*, 1890) (52)

Today the question of machine technology and machine civilisation is still capable of generating heated controversy. It is not, and never has been, one-way traffic in these disputes: there is an important voice which must be remembered, and in many important respects placed centrally to the discussion, which insists that the machine is of enormous human benefit. It is a voice, as likely as not, which will come 'from below' speaking a 'commonsense' language: domestic appliances such as electric washing machines are preferred to mangles and dolly-tubs; agricultural baling-machines are preferred to back-breaking pitch forks; mechanical excavators are better than picks and shovels; power-drills are preferred to handdrills. Even so, the steady erosion of craft skills and control over work, a process which has been accelerated by mass-production machine technology, continues to be a bitter issue for working people. 'Putting a spanner in the works' is one expression of the continuing and abrasive relationship between people and the factory (53). At a more formal level, opposition to machinery which displaces people from work by reducing the need for labour (container-ised transport systems, for example) remains a persistent question around which many trade union struggles are organised. The promise that automated production would open up new vistas of 'leisure' in the 'post-industrial' society has also come to be a bad joke, as working people find that either their working week is excessively long, or that they must occupy their leisure hours on the dole queue. Finally, there have been recurrent revitalisations of forms of Utopian and romantic thought which urge the need to return to a more simplified form of technology — whether it is William Morris's socialism which emphasised craft skill production and the social values of chivalry; 'socialism on a bicycle'; 'small is beautiful'; back-to-nature hippie solutions; or communard responses and innovations of all kinds.

These are difficult and complex movements, embodying a number of directions which are often in tension with each other. Quite commonly, the opposition to machine technology has become equated with an opposition to the material world itself; and in flight from the perceived horrors of materialism, reified science and machine civilisation, it lapses into subjectivism and mysticism (54). This can itself be either a pessimistic withdrawal from the world, or a joyous celebration of other creative human possibilities as in Edward Burne-Jones's Victorian exclamation: "The more materialistic Science becomes, the more angels I shall paint" (55). At other times, opposition to machinery and technology is little more than a lazy opposition to work itself.

Although it can, as for example in the traditions of Ruskin and Morris, be a means of placing the organisation of work and production centrally to an understanding of the creative possibilities of human life with the urgent necessity, however, to settle the question — men as tools, or men as men (56). Invariably associated with some aspect of romantic vision, invariably compounded by a nostalgic longing for what is past, these different and complex responses can nevertheless amount to more than a simple nostalgia for the old ways: and in an era which is becoming slowly conscious of the historical possibility of a disastrous ecological collapse on a planetary scale, these could be important and decisive social and cultural tendencies (57).

In fact, despite all the indications of the disappearance of opposition to technology after the 'teething troubles' of the industrial revolution, the last decade in particular has witnessed the re-emergence of powerful forms of resistance to the machine — principally to nuclear technology, to the massive threat of environmental pollution, and to new forms of mass communications systems — whether it is Tokyo's new airport, Concorde's sonic boom, or the urban motorway schemes which devastate city neighbourhoods. Characteristically, these new themes of resistance often remain locked into the old traditions of protest. Raymond Williams has remarked, for example, on the way in which resistance to 'urban renewal' schemes often echoes the preoccupations of earlier forms of resistance to 'improvement' and 'rationalisation':

> The convention of the country as a settled way of life disturbed by unwanted and external change has been complicated, in our own century, by very similar ideas about towns and cities . . . it is fascinating to hear some of the same phrases — destruction of local community, the driving out of small men, indifference to settled and customary ways — in the innumerable campaigns about the effects of redevelopment, urban planning, airport and motorway systems, in so many twentieth-century towns. . . I have heard a defence of Covent Garden, against plans for development, which repeated in almost every particular the defence of the commons in the period of parliamentary enclosures. (58)

Similarly, Japanese fishermen swinging axes to halt the building of a thermal electricity plant in a bay neighbouring their traditional fishing grounds phrase their opposition in terms which are dictated by localised problems — rather than through a more generalised opposition to machinery-as-such or to the forms of social consciousness which support such developments. Consequently, the message can only be: 'Go and build it near somebody else's traditional fishing grounds!' Again, resistance to the monster oil-tankers which bilge out their polluting filth into the sea provokes most public concern when it is tourist beaches which are threatened. Such resistance urges more 'rational' shipping routes, more 'rational' procedures for discharging oil cargo at sea, etc. — rather than through resistance to the petrol-driven private motor car, for example, and against the social and economic formations which depend on the mass ownership of the motor car. In other words, resistance tends to be both localised and superficial, in that what is demanded is a containment of a 'rational' level of pollution, or there is a polite request to foul someone else's patch, rather than striking out the root cause.

But what is the root cause? Is it technology as such? Or is it the political and economic formation which directs and dominates technological advance? Between those two poles of argument lies a bitter, and crucial, argument. Principally, the argument is between whether the solution is to somehow put technological advance into reverse gear, or whether a more humanly rational social order could guide the advance of technology without disastrous environmental damage.

It is becoming increasingly clear that 'pollution', and the development of ecological imbalances, is an inevitable consequence of an urban-industrial system which is based on massive industrial production, in particular the production of mass commodities which emphasises high mobility and high energy-consumption as a form of life; a form of life which is underwritten by a deformed Enlightenment philosophy which describes the relationship between 'man' and 'nature' as one of mastery rather than co-existence (59). The polar opposition — is it science *or* society which is at fault — proves to be a false dichotomy. The motor car, for example, is not simply a technical accomplishment; it is also a form of consciousness and a form of social relation, and technology as a whole should be understood as such an active social relationship, rather than as something which stands 'out there' as an objective limiting circumstance to the possibility of human progress. Understood as an active social relationship, there is no sense in which we can admit that the urban-industrial system is objectively doomed — however, deep and real the objective possibilities of ecological crisis may be — and the doom fantasists of impending planetary collapse are invariably simple-minded on this point (60). Equally, there is a common tendency to imagine that these problems of ecological imbalance are entirely new and unprecedented — a response which forgets, to take just one example, the profound ecological crisis faced by industrial cities as early as the 1840s at the time of Edwin Chadwick's excursions into smelly urban England and the so-called 'sanitary question' (61) — and which consequently fails to trace the connected history of such crises and to grasp the necessity of understanding them as an endemic feature of the urban-industrial systems.

The problem — both materially and intellectually — is undoubtedly aggravated in the postwar era. Materially, in the epoch which lies ahead, those endemic problems which flow from the aggressive exploitation of both human and natural resources are deepened by the objective limiting factor of global planetary resources as the urban-industrial social formation is generalised on a global scale. Intellectual discourse, on the other hand, is decisively altered by the way in which the traditional cry that 'science will provide' has come to have a distinctly hollow ring in the contemporary world — thus making the discourse at once less easy, but also potentially more fruitful in that it involves the de-mythologising of science. In particular, we can note that the organisation of scientific enterprise is so much more complete in the last quarter of the twentieth century that we can be consequently less hopeful that there is some new breakthrough, some new scientific miracle, lying just around the corner out of sight. The gurus of science, who pin their hopes on 'fusion energy', for example, as a way out of the energy crisis, recognise that several decades will be required to crack the production problems of this altered energy source — but they are understandably silent on whether it will be possible to maintain even a semblance of economic and political stability within the urban-industrial social formation during those decades which lie ahead, during which the scientific fraternity are beavering away for the 'way out'. Furthermore, all this — difficult enough as it already is — assumes that we can discount the recourse to war as a means of settling the so-called 'energy crisis' and access to raw materials and markets. The frighteningly spectacular developments in the technology of warfare in our century insist that a worried frown, or even downright pessimism, is the only appropriate human response to such an eventuality. "A smooth forehead", Bertolt Brecht suggested in one of his mottos, "betokens a hard heart".

If faith in the contribution of science to the resolution of these questions has dimmed, on the other side of the coin proposals to return to a more localised form of industrial production, to switch to a low-energy economy, etc., carry with them equally hollow Utopian assumptions: because, if it is technology itself which must be curbed and re-shaped, then how to accomplish this within a social formation which is predicated on the planned obsolescence of commodities, the constant up-dating of 'last year's model' by new and 'better' gimmicky commodities, high energy consumption, mass communications and high mobility, the circulation of money, etc.? The most likely scenario for the future, in fact, is that although resistance to technology will continue and develop new forms, it will prove unable to break out of its contradictions unless it can find a mature political expression. The history of resistance to the machine should tell us at least that much: because resistance to technology, even when it understands itself as resistance to technology-as-such, can invariably be seen to be attacking technological forms which are used *against* people rather than *for* them. It is thus a political and economic question which is contested through a form of inevitably nostalgic (although not necessarily nihilistic) violence.

I have raised a number of questions against the contemporary 'anti-technology movements' and 'ecology movements', although here I cannot hope, or even profess, to be able to answer them adequately. My aim in this essay has been more modest: namely, to describe and assess the historical significance of the kinds of nostalgic violence which are usually lumped together under the term 'Luddism'. One important historical reflection, as part of the wider reflection on the future direction of technological advance, must be of course "Were the machine-smashers right?" It is not such a silly question as it might at first seem. Or, at least, it is not silly if it is understood as something more than a request for a 'final' historical judgement. Because in trying to judge whether the machine-wreckers were right or wrong, we must remember once again that machine-smashing was a tactic which sometimes contested quite a number of different issues other than the machine itself — the level of wages, the availability of employment, control over work, the factory system of labour and factory discipline, political control, the stinking horrors of the emerging industrial cities, etc. Even if we only attend to those instances where opposition to machinery-as-such appears to have provided the guiding motive (and I have tried to argue that this kind of direct opposition, usually bound up with the decisive association between machinery and the factory systems of labour, has sometimes been played down by historians) then the answer to the question, "Right or wrong?", can only be approached by contrasting different and opposing schemes of rationality: principally, the rationality of rulers and capitalist manufacturers (in particular big capital), versus that of people who wished to defend a known way of life against the encroachment of the machine (which would include small capitalists as well as the common people). Then as now, the question of resistance to technology is dominated by these opposing rationalities — with the dominant rationality provided by that which is pledged to 'progress', 'advance' and 'improvement'. Furthermore, this dominant rationality, then as now, reserves for itself the title 'rational' and condemns resistance to the machine as a form of lunacy and irrationality.

In attempting to enquire into the motives of resistance to the dominant form of technology, it is necessary to confront deeply embedded cultural and political bans which deny any intelligibility to what is thus defined as a 'criminal nuisance' and 'vandalism' — a profound line of historical continuity which connects our own historical time with the apparently remote historical conjuncture of the 'pre-industrial' machine-smashers. In their day also, the powerful ruling elites of landowners and factory owners afforded supreme rationality to their own actions, and were blind to any others. Who could object to the building of the rational factory? Who could object to the dissemination of the rational machine? Who could object to the domination of rational market principles which placed hard cash before people? Who could object to the rational enclosure and improvement of the land? Indeed, when great men built their great country houses, sometimes tearing down whole villages simply in order to improve the view, who could object? And the answer bounced back: only rogues, vagabonds, men tainted with criminal folly, hooligans and vandals.

In what must figure as one of the earliest uses of the word 'vandal' in its modern usage, for shrugging off and slapping down the inherent social meanings of indifference to property, the *Manchester Mercury* in 1812 compared a huge crowd who attacked a cotton mill involving loss of life on their part, with "the very Goths and Vandals of antiquity" (62). It is no longer customary to drag the Goths into the act, but this early use of the term 'vandal' — a word which has now become wholly transformed and is usually reserved for application to the 'senseless' destruction of property by young people — carried with it by association a powerful imagery of barbarian hordes breaking into the precincts of civilisation. In that sense, nothing much has changed. When faced with the sometimes desperate energies of those who oppose the 'fast reactors', or the slow hiss of nerve gas, the culture which has thrown up these forms of civilised technology becomes equally self-satisfied and witless.

The historical fate of the machine-smashers is an unhappy one. History is written by the side which wins: or so it is said. And although socialist historians and historians 'from below' have unearthed a different historical narrative, public understanding of the history of machine-wrecking has been largely constructed by the winning side; that is, by the machine civilisation which has won the day and proved the machine-smashers wrong. However, the historical clock has a long spring and, to my own satisfaction at least, the historical fate of the machine-smashers has not yet been finally settled. Maybe, who knows, we have yet to see the last of machine-breaking. As the industrial world moves towards the historical possibility of a damaging ecological crisis, the machine could come to be understood once more as the enemy of human society — or, at least, people might form this impression of certain kinds of machines, such as those which were traditionally described as 'unfair' machines. So that, if the prophets of ecological imbalance and collapse are even remotely correct, then the day will come when men and women will attack the machines again. They will, of course, be condemned by public authorities as 'mindless hooligans' and as enemies of civilised progress — just as the Luddites and the others were condemned in their time. Their motives will certainly be different from those of the power-loom breakers and the men and women who smashed the spinning jennies and the factory clocks and who burnt down the mills. They will also require a technology of their own — that is, a technology of destruction more sophisticated than brute force if they are to tackle the elemental dangers of nuclear fission or germ warfare; for the improvised sledge-hammers and pikes which broke the jennies will not be equal to their task. Nevertheless we can assume that although their motives and methods will be different, their actions will be as equally intelligible as those of the 'goths and vandals' who went before them. Right or wrong, who knows, it may come to pass — when our culture and its preoccupations are as dead and unthinkable as the Luddites sometimes seem in our time — that the machine-breakers will have the last word.

(Some of the arguments in this essay have also appeared as "Goten und Vandalen — Verbrechen in historischer Perspektive", *Kriminologische Journal*, Vol. 9, no. 4, 1977; and "Goths and Vandals, Crime in History", *Contemporary Crises*, Vol. 2, no. 2, 1978; as well as forming the basis of a chapter in a forthcoming book, *Hooligan Terror: A History of Respectable Fears* to be published by Macmillan.)

Notes and References

1. I am not drawing here on all the work of these authors. The most relevant writings are: E.P. Thompson, 'Time, Work-Discipline and Industrial Capitalism', *Past and Present* 38 (1967) 56—97; E.P. Thompson, *The Making of the English Working Class*, Penguin, Harmondsworth, 1968; E.P. Thompson, 'The Moral Economy of the English Crowd in the Eighteenth Century', *Past and Present* 50, (1971) 76—136; E.P. Thompson, *Whigs and Hunters,* Allen Lane, London, 1975; E.P. Thompson, 'The Crime of Anonymity', in D. Hay *et al., Albion's Fatal Tree*, Allen Lane, London, 1975; E.J. Hobsbawm, *Labouring Men*, Weidenfield and Nicolson, London, 2964; E.J. Hobsbawm, *Industry and Empire*, Penguin, Harmondsworth, 1969; E.J. Hobsbawm, *Primitive Rebels,* Manchester University Press, Manchester, 3rd edn., 1971; E.J. Hobsbawm, *Bandits*, Penguin, Harmondsworth, 1972; E.J. Hobsbawm and G. Rudé, *Captain Swing*, Penguin, Harmondsworth, 1973; C. Hill, *The World Trend Upside Down*, Penguin, Harmondsworth, 1975; R. Williams, *Culture and Society 1780—1950*, Penguin, Harmondsworth, 1961; R. Williams, *The Country and the City*, Chatto and Windus, London, 1973; R. Williams, *Keywords*, Fontana, London, 1976.

    My text and argument relies so heavily on these works that I have not made cumbersome footnotes and references at every point, except in the case of direct quotations. There are other relevant works. One crucially important piece of historical research on the early industrial period is J. Foster, *Class Struggle and the Industrial Revolution*, London, Weidenfield and Nicolson, 1974, Also, although some of their interpretations are now disputed (see Note 17) there is the classical and unprecedented labour history of the Hammonds, in particular, J.L. Hammond and B. Hammond, *The Skilled Labourer 1760—1832,* Longman, Green and Co., London, 1919. Mention should also be made of a number of other important contributions to a history 'from below': the work of the History Workshop, founded at Ruskin College, Oxford under the direction of Raphael Samuel and represented in their journal *History Workshop*; G. Rudé, *The Crowd in History*, Wiley, London, 1964; G. Rudé, *Paris and London in the Eighteenth Century: Studies in Popular Protest,* Fontana, London, 1970; and, finally, G. Lefebvre, *The Great Fear of 1789,* New Left Books, London, 1974. It was, I believe, George Lefebvre who coined the phrase 'from below'.

2. For a useful account of how Manchester appeared to these distinguished visitors, see S. Marcus, *Engels, Manchester and the Working Class*, London, Weidenfield and Nicolson, 1974 in the chapter entitled 'The Town'. The quotation, below, from Sir Charles James Napier is taken from this source, p.46.

3. See Hobsbawm and Rudé, *op. cit.*, 1973, (Note 1).

4. See G. Pearson, *The Deviant Imagination*, Macmillan, London, 1975, Chapter 6.
5. N.J. Smelser, *Social Change in the Industrial Revolution*, London, Routledge and Kegan Paul, 1959, pp.227, 246.
6. H.J. Plumb, *England in the Eighteenth Century*, Penguin, Harmondsworth, 1950, p.79.
7. Thompson, *op. cit.*, 1968, (Note 1), Chapter 9.
8. *ibid.*, p.297.
9. Plumb, *op. cit.*, 1950, (Note 6), p.89. It is important to note a further level of contradiction in Plumb's account of the emergence of factory discipline. The concept of 'prison' to which Plumb is here appealing — a mass of men brought together, but nevertheless characteristically isolated, for the purpose of the regimentation of habits by the strict routines of time, discipline, containment, surveillance, etc., characteristics which are embodied in the 'barrack-like' architecture of prisons — is a concept of discipline and rehabilitation which does not actually belong to the eighteenth century, since the prison was only to emerge in this form as a dominant mode of social discipline and rehabilitation as part of the overall process of industrialisation and the creation of a factory system. Thus, it is in all its essential characteristics a concept of 'prison discipline' which only emerged in its developed form *after* the factory — with its own stringent disciplines of time, containment, isolation, surveillance, etc. — had become an established form of social organisation. In an important sense, the factory provided a model for the prison, rather than vice versa, and the discussion of the 'cellular system' in prisons for example is pre-dated by systems of surveillance and isolation in factories where fines were imposed on workpeople for talking, singing, swearing and even whistling! The interplay between these related material practices of discipline, and forms of social organisation and architecture, which are embodied in 'prison' and 'factory' are discussed at some length by Michel Foucault, *Discipline and Punish*, Allen Lane, London, 1977. The impact of the factory system of labour ön pre-industrial cultures was all the more decisive and terrifying in so far as factory discipline anticipated, in many important respects, the forbidding discipline of the prison as the dominant form of judicial control — a dimension which is quite lost in Plumb's account. See also: A.T. Scull, 'Madness and Segregative Control: The Rise of the Insane Asylum', *Social Problems* **24**, (3) (1977) 337–51; and Dario Melossi.
10. Plumb, *op. cit.*, 1950, p.150. (Note 6.)
11. D. Bythell, *The Handloom Weavers*, Cambridge University Press, Cambridge, 1969, pp.198, 199.
12. Thompson, *op. cit.*, 1967, (Note 1.)
13. Williams, *op. cit.*, 1976, pp.61–2. (Note 1.)
14. *Report of the Select Committee on Commons Inclosure*, Vol. 5, Her Majesty's Stationery Office, London, 1844, p.364.
15. S.G. Checkland and E.O.A. Checkland (Eds.), *The Poor Law Report of 1834*, Penguin, Harmondsworth, 1974. For example:
 The allotment of larger portions of land than ten rods to an individual, has this evil — if the labourer cultivates it himself with only the aid of his family, he overforces his strength, and brings to his employer's labour a body exhausted by his struggle. *Ibid.* p.280.
 Nor is this the only evil of the large allotments; a hovel perhaps is erected on the

land, and marriage and children follow. In a few years more, the new generation will want land, and demand will follow demand, until a cottier population similar to that of Ireland is spread over the country, and misery and pauperism are everywhere increased. *Ibid.,* p.280.

A farmer of the parish of Guildsfield in Montgomeryshire, stated that a labourer could not do justice to his master and the land if he had more than half an acre... He added that if he wanted a labourer, and two men, equally strong and equally skilful, were to apply, one of whom had a quarter of an acre, and the other one or two acres of land, he should without hesitation prefer the former. *Ibid.,* p.282.

16. 1826 was a bad year for utilitarianism. In the same year that the textile workers were rebelling against the material embodiment of utilitarian discipline and philosophy, John Stuart Mill, at the age of twenty, went mad. In his psychic revolt against the utilitarian discipline of his father's educational system, Mill experienced himself as a machine with all feeling drained away from him. See J.S. Mill, *Autobiography*, Oxford University Press, London, Chapter 5. Mill writes that he was cured by reading Wordsworth's poems which he describes as "a medicine for my state of mind", *ibid.,* p.89. In a strange premonition of romantic Laingian anti-psychiatry, Mill describes his earlier life under the influence of his father, and then writes of his release through madness: "The time came when I awakened from this [that is, his earlier life] as from a dream", *ibid.,* p.80. Elsewhere, in a letter to Carlyle, he described his "illness" in the following terms: "that period of *recovery* after the petrification of a narrow philosophy" (original emphasis); and he also wrote of "my former character, the character I am now throwing off". See F.E. Mineka (Ed.), *The Earlier Letters of John Stuart Mill,* Routledge and Kegan Paul, London, 1963, pp.181, 183. We can only guess at what untold madnesses were produced in the minds of common people by the new forms of discipline, rationality and containment. John Stuart Mill's case demonstrates that even the mighty were not exempt from the ill-effects of the emergent petrifying life forms.

17. It is difficult, in relation to this and other outbreaks of machine-smashing, to fully understand the level of organisation of the rioters. It is possible, for example, that the Luddite outbreaks of 1811 to 1813 moved in the direction of an organisation for a national uprising of the English people. The case is most forcibly argued in Thompson, *op. cit.,* 1968, Chapter 14 (Note 1) and *passim.* One major difficulty is that men who are plotting insurrection leave few traces, especially at a time when labour organisations are legally repressed as they were at that time in England. Most of the evidence for conspiracy, consequently, comes from government spies, informers and *agents provocateurs*; and it is sometimes argued that spies exaggerated the conspiracies in order to impress their paymasters. This is the line of argument in Hammond and Hammond, *op. cit.,* 1919, (Note 1), where the idea of an insurrectionary current amongst the common people is dismissed as the product of spies and informers with lurid imaginations. Thompson answers most of these points coherently and, if he is right, then the Luddite disturbances must certainly be placed in a much broader context of revolutionary agitation than is customary. The case is not settled to everyone's satisfaction, however. Compare, for example, M.I. Thomis, *The Luddites,* David and Charles, Newton Abbott, 1970, who argues, although somewhat unconvincingly, against Thompson. There is a discussion of this

conflict in F.K. Donnelly, 'Ideology and Early Working Class History: Edward Thompson and his Critics', *Social History* 2, (1976) 219–238.

As far as the 1826 outbreak is concerned, matters are even less clear. Traditional accounts describe a 'mob' travelling from town to town, wrecking looms and mills. If so, it would have had to be a particularly energetic mob. Imagine the energies of Samson which would have been required to destroy more than a thousand looms and several whole mills in just three days, to have successfully evaded troops throughout this period, and all this involving journeys between towns across rough moorland and hills carrying the equipment necessary for loom-breaking. The whole idea is most unlikely. What is at least equally likely is that the insurrection against the power-looms had been planned beforehand; there had certainly been rumours in the air for a few weeks before the first attack took place. For example, a week before the insurrection proper, a country calico weaver, William Varley of Higham, recorded in his diary: "There is a great disturbance at Accrington; they break the windows where the steam looms are; the country is all of an uproar for the poor weaver has neither work nor bread". Quoted in W. Bennett, *The History of Burnley 1650–1850,* Burnley Corporation, Burnley, 1948, Appendix. A third possibility is that it was a fairly spontaneous affair in which people from one town heard that the mills had been attacked in neighbouring towns, thought it sounded like a good idea, and decided to have a go themselves. What is certain is that the public authorities did not treat the matter lightly. Accounts vary, but several dozen people were charged, about three dozen were imprisoned, and ten men and women were sentenced to hanging, their sentences later commuted to transportation for life. As a response to the riots, turnpike roads were built in the area for easier troop movements, garrisons and prisons were built, and in one instance a factory was defended by cannon and a moat! Whatever the final outcome of this puzzle (if there is a final outcome) the recent research by John Foster, *op. cit.,* 1974, (Note 1), on working class organisation in the cotton belt reminds us that we should not underestimate the level or strength of the English working class in the early nineteenth century.

18. Quoted in C. Aspin, *Lancashire, The First Industrial Society*, Helmshore Local History Society, Helmshore, 1969, p.48.
19. Benjamin Hargreaves, *Recollections of Broad Oak,* Bowker, Accrington, 1882, p.42.
20. See, for example, the letter quoted by Thompson, *op. cit.,* 1968, (Note 1), pp.653–654: "We know that every machine for the abridgement of human labour is a blessing to the great family of which we are a part". The letter was written in May, 1812, at the point when Thompson suggests that Luddism was giving way to revolutionary organisation. It was signed 'Tom Paine'.
21. Ibid., p.648.
22. Quoted in Thompson, 'The Crime of Anonymity', *op. cit.,* 1975, (Note 1), p.281.
23. For 'wrecking', see J.G. Rule, 'Wrecking and Coastal Plunder', in Hay *et al., op. cit.,* 1975, (Note 1), pp.167–188. For smuggling, see C. Winslow, 'Sussex Smugglers', in Hay *et al., op. cit.,* 1975, (Note 1), pp.119–166. For smuggling in France, O.H. Hufton, *The Poor of Eighteenth-Century France,* Oxford University Press, London, 1974; and also Lefebvre, *op. cit.,* 1974, (Note 1).
24. Thompson, *Whigs and Hunters, op. cit.,* 1975, (Note 1) offers an extended analysis

of the Black Act. There is also a discussion of poaching and the game laws by Douglas Hay, 'Poaching and the Game Laws on Channock Chase' in Hay *et al., op. cit.,* 1975, (Note 1), pp.189–253. This criminalisation of the customary rights of the poor is, of course, reminiscent of Marx's early essay on the debates of the Rhenish Assembly concerning the theft of wood: K. Marx, 'Debates on the Law on Thefts of Wood', in *Marx and Engels Collected Works,* Vol. 1, Lawrence and Wishart, London, 1975, pp.224–263. Marx also promised articles on poaching and the game laws which never appeared. For a discussion of the relationship between the work of Edward Thompson and others on eighteenth-century crime and Marx's discussion of these questions, see G. Pearson, 'Eighteenth-Century English Criminal Law', *British Journal of Law and Society* **3**, (1) (1976) 115–131. For a discussion of Marx and the question of wood theft as it related to the situation in Germany, setting the matter in the context of political economy, see P. Linebaugh, 'Karl Marx, the Theft of Wood, and Working Class Composition', *Crime and Social Justice,* Fall-Winter 1976, pp.5–16.
25. Thompson, *Whigs and Hunters, op, cit.,* 1975, (Note 1), p.40.
26. *Ibid.,* p.131.
27. However, the actual number of convictions did not rise correspondingly and capital sentences were frequently commuted. For a brilliant analysis of how the 'rule of law' in eighteenth-century England operated through a blend of terror and mercy, see Douglas Hay's essay, 'Property, Authority and the Rule of Law', in Hay *et al., op. cit.,* 1975, (Note 1), pp.17–63.
28. Thompson, *Whigs and Hunters, op. cit.,* 1975, (Note 1), p.241.
29. W. Cobbett, *Rural Rides,* Vol. 1, Dent, London, 1912, p.17.
30. Thompson, *Whigs and Hunters, op. cit.,* 1975, (Note 1), pp.108, 113, 114.
31. Thompson, *op. cit.,* 1971, (Note 1); and also Thompson, *op. cit.,* 1968, (Note 1), Chapter 3.
32. Thompson, *op. cit.,* 1971, (Note 1), p.102.
33. For example: E.P. Thompson, ' "Rough Music": Le Charivari Anglais', *Annales Economies Societies Civilisation* **27**, (2) (1972) 285–312; N.Z. Davis, 'The Reasons of Misrule: Youth Groups and Charivaris in Sixteenth-Century France', *Past and Present,* No. 50, (1971) pp.41–75; Hobsbawn and Rude, *op. cit.,* 1973, (Note 1), p.39; J.R. Gillis, *Youth and History,* Academic Press, New York, 1974, pp.26–34.
34. Thomas Hardy, *The Mayor of Casterbridge,* Macmillan, London, 1974, p.302.
35. D. Williams, *The Rebecca Riots,* University of Wales Press, Cardiff, 1955.
36. Gillis, *op. cit.,* 1974, (Note 33), p.33. See also Thompson, *op. cit.,* 1971, (Note 1), pp.103ff.
37. Thompson, *op. cit.,* 1971, (Note 1), p.126.
38. Thompson, 'The Crime of Anonymity', *op. cit.,* 1975, (Note 1).
39. P. Linebaugh, 'The Tyburn Riot Against the Surgeons', in Hay *et al., op. cit.,* 1975, (Note 1) pp.65–117.
40. Quoted in Thompson, 'The Crime of Anonymity', *op. cit.,* 1975, (Note 1), p.330.
41. *Ibid.,* p.325.
42. This statement, taken from a meeting on the moors outside one of the North-East Lancashire cotton towns in 1842, is quoted in T. Newbigging, *History of the Forest of Rossendale,* Rossendale Free Press, Rawtenstall, 1893, p.325. It is estimated that there were 26,000 people present at the meeting, which might indicate that

such threats demanded that they were taken seriously.
43. Thompson, 'The Crime of Anonymity', *op. cit.,* 1975, (Note 1), p.271.
44. Marcus, *op. cit.,* 1974, p.4. What is not usually noted about the 1826 power-loom smashings is that they occurred in small, outlying towns which in a number of respects had not yet been drastically altered by the industrial revolution. Towns, that is, whose own population explosion belonged to the later period of Victorian buoyancy, from the 1850s through to the eventual downturn in the British textile industry in the 1920s. Therefore, the rioters were, in many ways, still essentially 'country people'. Crucially, however, they were country people who lived only twenty miles away from Manchester which was passing through a period of growth more explosive and catastrophic than anything to be witnessed by Engels in the 1840s. The experience of watching and listening, from the sleepy hollows of the Pennine towns, while Manchester — its steam-driven factories, its chimneys, and its filthy slums — grew at such an alarming rate, provides an important inferential clue for reconstructing the motives of the great power-loom smashing of 1826. Significantly, the riots had few repercussions in the more developed centres of population and industrialisation, and although there was a brief skirmish in Manchester a day or so later when power-looms and mills were attacked in some quarters, the trouble was easily contained. (See Hammond and Hammond, *op. cit.,* 1919, (Note 1), pp.127—8.) The origins of the 1826 riots in the small towns, as opposed to the city, reminds us once again that a simple economic determinism which puts the insurrection down to 'slump' provides an insufficient argument.
45. C.R. Fay, *Life and Labour in the Nineteenth Century*, Cambridge University Press, Cambridge, 1920, p.178.
46. Bennett, *op. cit.,* 1948, (Note 17), p.292.
47. W. Cooke Taylor, *Notes of a Tour of the Manufacturing Districts of Lancashire,* Frank Cass, London. 1968 edn., p.68.
48. *Ibid.,* pp.68—69.
49. *Ibid.,* p.69.
50. *Ibid.,* p.91.
51. K. Marx and F. Engels, *Manifesto of the Communist Party,* Foreign Languages Publishing House, Moscow, n.d., p.56.
52. William Morris, *Three Works*, Lawrence and Wishart, London, 1973, p.295.
53. See, for example, the discussion of various interruptions of the work process, in L. Taylor and P. Walton, 'Industrial Sabotage: Motives and Meanings'. In S. Cohen (Ed.), *Images of Deviance*, Penguin, Harmondsworth, 1971, pp.219—245.
54. For example, T. Roszak, *Where the Wasteland Ends,* Faber, London, 1972.
55. Quoted in E.P. Thompson, *William Morris: Romantic to Revolutionary,* Lawrence and Wishart, London, 1955, p.86.
56. For example, the delightful essay by William Morris, 'A Factory As It Might Be' (1884) reprinted in B. Simon (Ed.), *The Radical Tradition in Education in Britain,* Lawrence and Wishart, London, 1972, pp.291—299. *News From Nowhere* offers a more sustained account of Morris's vision of a socialist future based on craft production, in Morris, *op. cit.,* 1973, (Note 52). There is a detailed and illuminating discussion of Morris, together with Ruskin's influence, in E.P. Thompson, *William Morris: Romantic to Revolutionary,* Merlin, London, 2nd edn., 1977; also, E.P. Thompson, 'Romanticism, Moralism and Utopianism', *New Left Review,* no. 99,

(1976) 83–111.
57. See H.M. Enzensberger, 'Critique of Political Ecology', *New Left Review,* no. 84 (1974) 3–31; Thompson, *op. cit.,* 1976, (Note 56).
58. Williams, *op. cit.,* 1973, (Note 1), pp.291–292.
59. For example, M. Horkheimer and T.W. Adorno, *Dialectic of Enlightenment,* Allen Lane, London, 1973; and M. Horkheimer, *Critical Theory,* New York: Herder and Herder, New York, 1972.
60. See Enzensberger, *op. cit.,* 1974, (Note 57).
61. E. Chadwick, *Report . . . on an Inquiry into the Sanitary Condition of the Labouring Population of Great Britain,* Her Majesty's Stationery Office, London, 1842. For the political implications of Chadwick's treatment of the 'sanitary question', see Pearson, *op. cit.,* 1975, (Note 4), pp.160–173.
62. Quoted in Bythell, *op. cit.,* 1969, (Note 11) p.198.

IS ANTI-SCIENCE not-SCIENCE?
The Case of Parapsychology

T.J. PINCH and H.M. COLLINS
University of Bath

Introduction

In this paper, we try to show that it is not possible to ascribe, unambiguously, the label 'anti-scientific' to a set of ideas. Anti-science is a description applicable to certain human activities; it is not a quality of certain ideas. We will illustrate this by looking at the case of parapsychological ideas (1). This approach implies that we do not endorse any positive definition of anti-science within this paper. The argument is intended to show the inappropriateness of definitions that would identify anti-science with sets of ideas as opposed to sets of historically specific actions. We will, however, look at some of the reasons that are given when ideas are so labelled, and we will put forward suggestions as to why actors should be interested in applying the label in this way to parapsychological ideas.

Much of the recent wave of what has been called anti-scientific thought has included a component seemingly associated with parapsychology. For instance, recent editorials in *Science* and *Nature* (2) attacking the growth of pseudoscience include parapsychological phenomena along with the more 'fringe' beliefs of Velikovsky, von Daniken and the like. *The Humanist* and its British equivalent, the *New Humanist,* have run articles debunking parapsychology in the context of attacking the growth of occultism and interest in anti-science (3). The association of parapsychology with this broader spectrum of interests can be seen in the following quotes concerning the rise of anti-science:

We can think of psi, for example, as something which is irretrievably beyond the scope of science, at least any *science* besides parapsychology. Psi in this sense, is not only extra-scientific, it is supernatural and anti-scientific (4).

How can man ever hope to solve the problems of his existence on this planet if, in crisis, he seeks comfort and guidance from the mystical and occult rather than stare reality in the face?
   For this, I believe, is what is happening to Western Society in its present state of turbulence. Where science and technology fail, superstition steps in as a welcome house guest. When we look at our world and see only a cruel present and a soulless future, then ghosts must walk and metal bend to comfort us. When politicians and philosophers fail to guide and console, the time of the psychic Messiah is nigh (5).

It would have amazed the Victorian steadfasts of science how confused some of our attitudes towards science still are. Instead of the logical world they hoped for and tried to work in, there is a discernible tendency for the public and even some practioners of science to turn their backs on science and become proccupied with the bizarre and the magical.
   Mr Uri Geller is only the most recent to cast doubt in the public mind on the efficacy of rational explanation (6).

The disquiet over the growth of the 'pseudosciences' has even led to the formation of a 'Committee for the Scientific Investigations of Claims of the Paranormal'. The newly formed committee claims to be scientifically neutral — as the co-chairman Paul Kurtz put it:

We wish to make clear that the purpose of the Committee is not to reject on a priori grounds, antecedent to inquiry any or all (paranormal) claims, but rather to examine them openly, completely, objectively, and carefully (7).

However, the first two issues of its journal, *The Zetetic*, have featured mainly articles critical of the pseudosciences including, of course, aspects of parapsychology. And, Paul Kurtz, (now chairman of the committee), perhaps in a less guarded comment, is quoted as follows in an article in *Science News:*

Often ..... the least shred of evidence for these claims is blown out of proportion and presented as 'scientific' proof. Many individuals now believe that there is considerable need to organize some strategy of refutation. Perhaps we ought not to assume that the scientific enlightenment will continue indefinitely; for all we know, like the Hellenic civilization, it may be overwhelmed by irrationalism, subjectivism and obscurantism (8).

What is more, Marcello Truzzi, the other founding co-chairman of the committee, has recently resigned because, in part, of the lack of scientific 'objectivity' shown in its activities.

The reaction of bodies such as the 'Committee' has then been directed at what has been perceived as a reaction to science encouraged by belief in the paranormal, in turn encouraged by parapsychologists' attempts to gain scientific acceptance for their claimed findings. Yet this very reaction draws attention to the point that this paper is concerned with. One group of actors perceives the set of ideas associated with parapsychology to be perfectly compatible with science and attempts to seek scientific recognition for them, whilst another group of actors perceives the ideas to be incompatible with science and attempts to disarm the first group by attacks *from science* on them. The situation in actuality is even more complicated because a third group of actors also perceive the ideas to be incompatible with science yet are content to seek recognition for the ideas independently of the institutions of science. These different modes and forums of attack, and strategies for the acceptance of ideas have been discussed in an earlier paper (9) but it seems clear that this variety of activities in itself indicates that a set of ideas does not necessarily entail a set of activities which can unambiguously be described as 'anti-science'.

It seems that *ideas* may be called 'anti-scientific' for three main reasons. Firstly, they may be called 'anti-scientific' because they do not meet some epistemological maxim for the constitution of legitimate scientific knowledge. Such maxims might be, for example, a version of the hypothetico-deductive model, the principle of falsification or perhaps some version of inductive logic. A recent example of the deployment of an epistemological maxim used for the purpose of detecting anti-science is Sachs's claim that such ideas 'do not actually entail cause-effect relations or the rigorous hypothetico-deductive method that is required of *scientific explanations* (10). However, Sachs uses this maxim to reject astrology, black holes, the twin paradox of relativity theory and the Copenhagen Interpretation in quantum mechanics along with parapsychology. In revealing the lack of discrimination of this demarcating criterion Sachs himself seems to reveal its unworkability.

Secondly, ideas might be thought to be part of anti-science because they have not been constituted according to some canonical methodology for establishing scientific knowledge. For instance, it is frequently claimed that phenomena must be embedded within theories and must engender repeatable experiments in order to be a part of science. In a previous paper (11), we have documented some of these attempts to reject the ideas of parapsychology because of such factors. We found that, in general, this method of casting doubt on the scientific status of parapsychology was not as convincing in practice as some authors have claimed.

The third reason that can be given for placing ideas within the category of anti-science is that the *content* of these ideas is contradictory or incompatible with established scientific knowledge. It is this latter means of seemingly rendering ideas anti-scientific that we are concerned with in this paper. In other words, it is not the epistemological status of the ideas or their methodological constitution which concerns us; it is the claim that the ideas a priori are incompatible with science.

Parapsychology is perhaps unique amongst activities that have been associated with today's anti-science movement in that it has a long history of struggle to gain scientific acceptance. Since Rhine's work of the 1930's, it has been in a relation of what might be called 'institutionalised conflict' with science. The question of parapsychology's scientific status has engaged the attention of parapsychologists, orthodox scientists and philosophers and precipitated a body of argumentation concerning the compatibility of parapsychological ideas with those of science.

The particular importance of this debate arises, as we shall see below, because some arguers, who have claimed that there is a conflict between the two sets of ideas, have gone on to dismiss the ideas of parapsychology *because* of this incompatibility. The most refined version of this dismissal is Hume's argument concerning miracles which states that:

A miracle is a violation of the laws of nature; and as a firm and unalterable experience has established these laws, the proof against a miracle, from the very nature of the fact, is as entire as any argument from experience can possibly be imagined ... (12).

Thus, because parapsychology is incompatible with science, the evidence on which science rests must also count against parapsychology. But in order to make such an argument work it must first be established that the two sets of ideas are incompatible.

We will look at the argument over the incompatibility of the two sets of ideas here in order to suggest that its outcome is not likely to be determinate. If this is true, then it is certainly true that a priori it cannot be said that parapsychological ideas conflict with science, and a fortiori it cannot be said a priori that they are anti-scientific. We will look, therefore, at the question of whether parapsychology and science contradict each other by focussing on this compatibility debate.

Arguments About Compatibility

To begin to think about what might be meant by saying that the content of two sorts of endeavour are in contradiction is to realise how unclear such a notion is. The only things that philosophers can be sure do contradict (with reservations, of course) are, p and *not p*, but to show that elements within scientific endeavours are related like p and *not p* must be attended by difficulties concerning disagreements between parties over the correct translation of p and *not p* into other symbolic universes. Do all parties mean the same thing by, for example, telepathy and precognition, and science and causality? Again, would philosophers agree about the fundamental elements within scientific systems that need to be compared in order to establish a broad contradictory relationship? One would imagine that they would not. For instance, would not a Kantian attempt to demonstrate contradictions at the level of the a priori perceptual categories that make available the world of phenomena? In such a scheme a Kantian might rule out precognition by virtue of its conflict with categories related to Newtonian time and causality. On the other hand, a logical empiricist would presumably be much more concerned with a conflict of potential experimental evidence, refusing to accept that the manipulation of logical symbols other than those of an observation language could yield any new knowledge about the existence or otherwise of phenomena. Students of the history of philosophy, could, no doubt, extend this discussion further, but we will adjourn it in favour of an examination of arguments that have actually been put forward in order to make one sort of claim or another about the relationship between parapsychology and other scientific work.

Arguers, on the whole, do not seem unduly exercised by philosophical principles. They do not seem to worry, for example, about the analystic nature of the notion of contradiction, or logical reasoning in general. Thus some critics of parapsychology, as we pointed out above, argue that psi-phenomena cannot exist because of their incompatibility with science. For instance, G.R. Price, a well known sceptic, sets out to show the improbability of the existence of the paranormal "..... by showing that ESP is incompatible with current scientific theory" (13). Similarly, C.E.M. Hansel, another leading sceptic, suggests: "In my view, a priori arguments determine our attitude towards an experiment, and may save time and effort in scrutinising every experiment ....." (14). Proponents of the paranormal find such arguments of sufficient merit to require replies in kind.

Views like these, and counter-arguments have been presented in the following formats:

(1) A belief in the unity of science (implicit) and the incompatibility of psi-phenomena with science or certain of its characteristics leads to the conclusion that psi-phenomena are spurious.

(2) A belief in the existence of psi-phenomena and the incompatibility of psi-phenomena with some part of science leads to the conclusion that science must be changed or undergo a revolution led by parapsychology and therefore psi-phenomena are important.

These two arguments have at their core the same claims regarding the incompatibility of psi and science though they are deployed by both sceptics and believers, respectively — hence some of the strange bedfellows that will be found in the later sections.

(3) Psi-phenomena are compatible with science so they are not impossible. This is one type of answer to argument 1.

(4) Science is open ended, or pluralistic so that even if psi phenomena are incompatible with science or parts of science this doesn't mean they are impossible.

This is another type of answer to argument 1.

(5) Science is full of incompatible pieces called anomalies. Psi-phenomena are like these and not very important.

(6) Psi-phenomena are incompatible with current science, as it is perceived, by definition. (This says nothing about their existence.)

(7) Everything that exists must be brought under the aegis of science and therefore psi-phenomena are part of science irrespective of any contradiction they seem to entail.

The majority of what follows consists of incompatibility claims generated by proponents of formats 1 and 2 and answered by proponents of format 3. All the other formats are represented too. We have taken a number of liberties with the arguments of writers, in particular we have lifted most of them out of context with scant regard for the justification which their authors would have given them. In the main body, we are not even concerned with the interests that authors thought their arguments were serving, though we point this out now and again. What we are trying to show is the essentially open-ended nature of the whole incompatibility debate, and so all we would claim about the argument 'fragments' we assemble, is that the beliefs they embody have been held by 'rational' people and put into print in 'responsible' forums. There is no attempt here to write a history of the debate, or a history of ideas. We are not concerned to do each argument justice but to try to suggest that there are enough tenable viewpoints, available to rational people to make an a priori decision on the compatability question unlikely, as either side can be defended by an appropriate shift of ground, or deployment of one kine of subsidiary argument or another. The attempt to subsume human *actions* under logical categories fails.

The arguments are arranged in subsections with claims of incompatibility coming first, and the arguments which avert this conclusion following on. One might imagine the arguments which propose incompatibility as translating something scientific and something parapsychological into p's and q's in such a way that $p \Rightarrow not\ q$ or $q \Rightarrow not\ p$ so that to hold p and q would be to hold p and $not\ p$. The next arguments show how a different translation avoids the contradiction. Nothing as hard or well defined as logical entailments are ever broached or suggested of course. We are looking at practical arguments in the spirit, we hope, of Toulmin (15).

The Case Of Parapsychology

1. Parapsychology Incompatible With Science?

At the highest level of abstraction we can find arguments, or argument fragments which claim simply that some or all parapsychological phenomena contradict the whole of scientific knowledge. This view is represented by Hansel for instance:

The whole body of scientific knowledge compels us to assume that such things as telepathy and clairvoyance are impossible (16).

Other arguers have claimed that parapsychological phenomena contradict certain basic principles pertaining to science as a whole. For example, G.R. Price writes:

But the conflict is at so fundamental a level as to be not so much with named 'laws' but rather with basic principles (17).

That the conflict is with immutable scientific knowledge is claimed by Willis:

The conclusions of modern science are reached by strict logical proof, based on the cumulative results of numerous ad hoc observations and experiments reported in reputable scientific journals and confirmed by other scientific investigators: then, and only then, can they be regarded as certain and decisively demonstrated. Once they have been finally established, any conjecture that conflicts with them, as all forms of so-called 'extra-sensory perception' plainly must, can be confidently dismissed without more ado (18).

These authors have been interested in establishing the non-existence of psi via an argument of type 1. Broad, Mundle, H.H. Price and Shewmaker and Berenda have argued a broadly similar thesis to a different end. Rather than using the contradiction to cast doubt on the reality of psi (19) they use it, instead, to question the status and general validity of the principles with which psi is held to conflict.

Broad defines several 'Basic Limiting Principles' of four main types. These are concerned with: causation, the relation of mind to matter, the dependence of mind on brain, and limitations on ways of acquiring knowledge. The scope of these principles is outlined by Broad:

.... I think that they will suffice as examples of important restrictive principles of very wide range, which are commonly accepted today by educated plain men and by scientists in Europe and America (20).

Board then goes on to show that certain paranormal phenomena are prima facie in contradiction with one or more of his principles.

Mundle adopts certain principles outlined by the philosopher Bertrand Russell for the analysis of science. In particular, Mundle claims psi-phenomena contradict a principle of causal connection, and he justifies the use of this principle by an appeal, similar to Broad's, that "most scientists seem confident that the principle is universal and necessary" (21).

If the contradiction between these principles and psi-phenomena is to be used to question the general validity of the principles then clearly this serves to emphasise the importance of psi-phenomena. For instance, H.H. Price writes:

> Psychical Research has succeeded in establishing various queer facts about the human mind, but (some people) think that these facts are mere curiosities and oddities of no particular importance . . . On the contrary these queer facts are not at all trivial, and it is right to make the greatest possible fuss about them. Their very queerness is what makes them so significant. We call them 'queer' just because they will not fit in with orthodox scientific ideas about the universe and man's place in it (22).

It is clear that H.H. Price is here concerned to draw attention to the significance of psi rather than its non-existence through the use of his arguments. Similarly, both Broad and Mundle are concerned with exploring the relationship of accepted (or at least hypothetically possible) psi-phenomena with science as it is currently understood, and therefore exploring any (hypothetically) necessary changes in such knowledge.

Shewmaker and Berenda make a similar point:

> These phenomena appear to violate principles which are basic to our entire scientific mode of explaining all other physical and psychological phenomena. Yet they, no less than any other events, demand explanation (23).

These attempts to establish the contradiction between parapsychology and science at the level of general principles are examples of, what we described earlier as argument formats 1 and 2 (see p. 226). However, these types of argument can be answered by an arguer who claims that psi and science do not entail any contradiction (format 3). One philosopher who has consistently argued for the compatibility of the two fields has been Michael Scriven. For instance, he finds such claimed contradictions between general principles of science and psi-phenomena to be insubstantial and a matter of philosophy rather than physics. He writes:

It is true that certain vague general principles which characterize many of our laws are rejected by ESP supporters, but I would class these general principles as being at the level of philosophical rather than physical insights, and consequently even more readily subject to reformulation (24).

One particular principle with which it has frequently been claimed psi-phenomena conflict is the doctrine of 'epiphenomenalism' and its close relative 'the materialistic conception of personality'. Hansel (a sceptic) and H.H. Price (a believer) again make strange bedfellows when it comes to their opinions on this principle. Hansel in suggesting principles of knowledge which conflict with ESP includes the epiphenomenalist notion that mental processes are dependant on physical processes in the nervous system of the person who experiences them. Hansel uses this as an a priori argument against psi-phenomena (25). Price finds the same contradiction, though he is using it to argue against the universal applicability of the materialist principle:

... if we consider the implications of telepathy, the most elementary and the best established phenomenon in the whole field of Psychical Research, we shall see that they are incompatible with the Materialistic conception of human personality (26).

Although Hansel is arguing against the existence of psi-phenomena by this principle (format 1) and Price is arguing against the validity of the principle (format 2) both again accept that there is a contradiction.

This contradiction between psi and science, established through the intervening variable of materialism, is defeasible too, however. Both Mundle and Ducasse think that it is possible that a way will be found to reconcile psi-phenomena with the materialist viewpoint. For instance, Mundle points out that current theories of physics may not be complete or correct and that materialism, even in current physics, allows for the explanation of certain interphenomena such as the 'force' of gravity. He sees no reason why explanations of ESP should not be found which do not challenge materialism. As he writes:

It may be argued that 'X cannot be explained by physics' does not entail 'X disproves materialism', unless 'materialism' means, among other things, that the current theories of physics are correct and complete, which no one would claim. . . . . 'Materialism' need not be defined so as to entail that physical things can interact only by contact or by means of physical radiations or fields. A materialist is free to maintain that the 'interphenomena' of physics (radiations, the 'force' of gravity, etc.) are logical constructs, and that the failure, so far, to construct inter-phenomena to explain ESP need not cause ontological discomfort (27).

Ducasse makes the similar point that there may be an, as yet, undiscovered part of physics, such as a sub-sub-atomic level, which might lead to the explanation of psi-phenomena but would not overthrow the materialist conception of science:

That matter may have sub-sub-atomic constituents, and that these might have properties capable of accounting for ESP and PK is, of course, at present pure speculation. I introduce it only to make clear that the reality of these and of other kinds of paranormal phenomena would not in principle require abandonment of a materialistic conception of the universe . . . (28).

These moves defeat the materialism/psi incompatibility (format 3) and open the way for psi to be compatible with a materialistic science.

Thus, in our survey of the literature, we have found nothing that would demonstrate definitively that the content of parapsychological ideas is in conflict with general scientific principles. Parapsychology does not seem to be *necessarily* anti-scientific for this reason.

2. Parapsychology Incompatible With Physics?

For some authors, there is no sense in making a distinction between the relationship of psi to science, and the relationship of psi to physics. For instance, Michael Scriven suggests that:

. . . physics is, in an important sense, *co-extensive* with science (29).

But to others (see below, p. 241), the distinction is important. Thus the following extracts of arguments which suggest that psi is incompatible with physics in particular (as we know it) might be significant:

. . . it is true that the evidence and conclusions of parapsychology . . . do not seem to fit into the panorama of physics today (30).

Psi and physics are irreconcilable in terms of what we know today, and present knowledge is the only knowledge we have at our disposal in trying to relate them (31).

Mundle makes the same point for physical mechanisms in general and with specific reference to clairvoyance and PK:

The relevance of psi phenomena is, of course, that it seems impossible in principle to explain these powers in terms of physical mechanisms . . . (32).

To explain clairvoyance, physics would have to undergo a major revolution (33).

Would any physicist agree that P.K. is compatible with what is thought to be established in physics? Most physicists would, I think, be inclined to take it as certain that a person's volition cannot be a *proximate* differential condition of physical events occurring at a distance from that person's body . . . (34).

These suggestions that there is a contradiction with *physics* are also defeasible. The definition of physics must be related to the definition of physical, and certain collections of arguments have shown that this definition is by no means unproblematical. For instance in the collection of papers *Science and ESP* (35), concerning the relation between parapsychology and science, several meanings of the word 'physical' emerge. For example, Mundle (36) claims physical means "visible and tangible", Price (37) claims that nothing is a physical event unless "it is perceptible by means of the sense-organs; either directly, or indirectly"; and, finally, Dobbs (38) claims that physical agencies are the kind "acceptable in principle to physics as currently practised". Also no clear meaning of the term 'physical' emerged from a debate in the *Journal of Parapsychology* on the issue of 'Physicality and Psi', with C.D. Broad (39) and Michael Scriven (40) both pointing out that it was not possible to given an agreed meaning to the term, and with Pratt (41) asserting that physics dealt with changes within the material universe that are capable of being described in terms of "time-space-mass-energy-relations". This definition was in turn questioned by Scriven (42) who pointed out that entropy must be included in any definition of physics: and hence degrees of order and disorder; hence information-content and correlations between messages, i.e., ESP phenomena. In other words, whether or not there is a contradiction between psi and physics depends on the definition of physicality adopted and no agreed definition has emerged from the debate.

Another way of removing incompatibilities between psi and physics is suggested by McConnell. He exploits the logical possibility entailed in the dual interpretation of any psi experiment (see below p. 239) to avoid the conclusion that PK (mind over matter) is incompatible with physics. By interpreting PK experiments in terms of clairvoyance (mind sensing the state of matter) it is possible to avoid breaking the cause and effect chain; a break which it is claimed means that PK is in contradiction to the notion of cause and effect in physics (i.e. it is non-physical):

However, the demonstration of the lack of these relationships must be exhaustive to be conclusive — a logically impossible task ... there would appear to be a logical difficulty ... in any attempt to show violation of causal sequence. If our wish follows the dice throw, is it PK or clairvoyance? This difficulty would seem to rule out for PK any positive test for non-physicality as we have defined that term (43).

It has also been argued that many of those wishing to press the case that psi and physics and incompatible have a restrictive view of physics. As Scriven puts it:

Now the interesting claim would be that there is some *impossibility* about the idea of physics emcompassing psi. But I see no possible way of justifying that claim ... the opposition to it [psi] is largely based on an absurdly parochial idea of the limitations of physics (44).

Murphy and Honorton make a similar point:

It is the physics of the 19th century, persisting in terms of current space-time patterns, that makes the phenomena 'impossible' (45).

The debate over the incompatibility of physics and ESP has been conducted almost exclusively within the framework of nineteenth-century deterministic physics, wherein the ultimate constituent of physical reality was still believed to be solid matter (46).

Burt, too stresses the amenability of contemporary physics to parapsychological notions:

... it would seem that there is nothing whatever in *contemporary* physics which would preclude the apparent anomalies presented by psi-phenomena (47).

Many of the unexpected phenomena encountered in the study of quantum physics would seem to be 'characterized by not conforming to physical laws' as ordinarily understood . . . (48).

Murphy and McConnell stress the 'open-ended' nature of physics (a format-4-type move):

Of course what the *physics of the future* may reveal none of us should be foolish enough to predict; this is perhaps one of the reasons for being a little hesitant about insisting so strongly today that we know we are dealing with psychic or spiritual factors rather than physical ones (49).

Perhaps at the sub-atomic level, physics borders on parapsychology in some not-yet-understood sense, so that the distinction between them cannot be clear-cut (50).

In these arguments, it is as though the question of the compatibility of psi with physics depends upon which 'historically frozen' moment of physics is selected for comparison with parapsychology. The final point is that as the history of physics will continue to unfold for ever, a final decision for incompatibility can never be reached.

Thus it seems that the nature of physics in general does not mean that parapsychology is necessarily in conflict with it, or necessarily 'anti-physics'.

3. Specific Physical Principles

It might be thought that the incompatibility of psi and science can be argued away more easily at the general levels of science and physics we have been looking at, but, when specific physical principles are considered, the room for argument is much less. It transpires that this is not the case.

The argument regarding incompatibility has been carried into discussion of such specific physical principles as, for instance, energy. Murphy makes the following comment regarding energy problems and psi-phenomena:

We know, as we do, with some degree of certainty that we are not dealing with any of the types of physical energy with which contemporary physics is concerned (51).

and Mundle (52) and Burt (53) suggest that energy resources in the brain are too small. The problems of space and time are mentioned by H.H. Price and McConnell:

> In telepathy, one mind effects another without any discoverable physical intermediary, and regardless of the spatial distance between their respective bodies (54).

> Specifically, physics is the expression of matter relationships in space-time co-ordinates. Thus the absence of any regular relationship between experimental phenomena and space or time should, by definition, be proof of non-physicality (55).

A similar point is made by G.R. Price (56).

Another problem put forward is the informational problem. For example, Mundle (57) and G.R. Price (58) — again, strange bedfellows — point out that a subject can obtain information paranormally from a specific target (e.g. a card in the middle of a pack) without interference from other physical objects in the surrounding area (e.g. the other cards in the pack). Beloff makes a similar point:

> ... the crux of the problem as I see it, lies, not so much in specifying what kind of energy might surmount spatial and temporal distances or material barriers, but rather in explaining how it comes about that the subject is able to discriminate the target from the infinite number of other objects in his environment (59).

A related point is made by Mundle (60) and G.R. Price (61) in remarking that physical barriers seem to be pervious to ESP. Finally, Burt (62) suggests that there is no structure in the human body capable of transmitting or receiving ESP signals.

Thus it is argued that all these physical principles are in conflict with psi-phenomena.

We will look at the way some of these specific points have been answered before taking up, at slightly greater length, another specific incompatibility claim, namely that psi-phenomena do not obey the inverse square law — thought to be typical of all physical forces.

With regard to the energy incompatibility, Margenau has suggested that, in some cases, parts of physics itself do not obey normal energy relationships. Thus:

At the forefront of current physical research, in the fields of quantum theory and elementary particle physics, the principle of conservation of energy is frequently breached because we find it necessary to invoke the existence of 'virtual processes' (63).

This has the effect of undermining the rhetorical power of any contradiction between psi-phenomena and such energy considerations.

Regarding the informational problem, Meehl and Scriven write:

> .... no simple radiation theory can explain the Pauli Principle and one can no more refute it by saying "How could one electron possibly know what the others are doing?" than one can refute the ESP experiments by saying "How could one possibly read a card from the middle of the pack without interference from those next to it?" (64)

Again this has the effect of undermining the incompatibility.

And, of course, regarding any of the other objections the open-ended nature of science may be cited just as before in the case of seemingly more abstract physical incompatibilities. Thus, Burt writes:

> The foregoing objections (in which Burt lists many of those listed above) ... would seem to completely rule out the ever-popular notion of 'brain radio'. They do not of themselves, however, exclude all other physical modes of transmission. The new theories of quantum physics may conceivably contain possibilities which have not yet been adequately explored (65).

Because it has not been demonstrated decisively that there are any specific physical principles that conflict with parapsychology it would seem that any claim that parapsychology is necessarily anti-scientific because it conflicts with certain physical principles is unwarrantable. We will now go on to consider a specific physical law with which, many arguers claim, parapsychology is in conflict.

4. Inverse Square Law

The inverse-square-law objection ties in with the apparent lack of attenuation of telepathic signals over distance. Most physical forces seem to be reduced in strength as the square of the distance from the source and it is expected that all physical forces should be reduced in the same way (66).

One answer to this suggestion is to produce a counter-example from physics itself, as do Margenau, Dobbs, and Rush:

... it should also be recalled that not all interactions obey an inverse square law — in fact almost none do .... An electric field in front of a charged plane of infinite extent shows no attenuation at all (67).

... the inverse square law does not necessarily apply, even to ordinary electro-magnetic radiations. For, in the case of radio waves, the strength of a signal as received may even increase with increase of distance between transmitter and receiver, due to ionospheric effects and various forms of 'ducting' (68).

... it must be borne in mind that the inverse-square propagation of energy is seldom realized in practice. Such effects as diffraction, reflection, refraction and absorption, as well as deliberate 'beaming' in the case of radio signals, modify the simple spatial distribution (69).

Essentially Rush is claiming here that if the radiation analogy is pushed far enough, we have to accept that psi radiation can be modified away from the ideal inverse square law in the same way as he claims for other physical radiations.

The advantages, in terms of avoiding the incompatibility, by pushing the radio analogy even further, are shown by Mundle, who writes:

ESP is being assimilated to radio transmission; radio sets have amplifiers and volume controls; we may suppose then that the brain possesses an automatic volume control which amplifies weak ESP signals (70).

This means that the lack of attenuation can be explained by large amplification in the brain for distant signals and small amplification or even reduction for near signals. So, in extending the analogy, Mundle is also removing the incompatibility by pointing out that psi can still obey an inverse square law.

A similar approach is adopted by Meehl and Scriven, who write:

... since we have no knowledge of the minimum effective signal strength for extrasensory perception, the original signal may well be enormously attenuated by distance and still function at long range (71).

So ESP may again obey an inverse square law over some of its range.

Another alternative is suggested by Burt, he writes:

> ... the transmitter's brain might include a mechanism which, like the laser, could produce amplification by stimulated emission of the relevant radiation. The radiation could then perhaps be concentrated and directed in almost linear fashion towards recipients ... (72).

A different sort of objection to the inverse square law incompatibility is raised by Rush when he considers the information content of the assumed radiation:

> ESP is the perception of a *signal*, i.e. of a systematic pattern impressed on the assumed radiation, much as audible speech frequencies are impressed on a radio wave. The mind interprets the *relative* intensities which compose this pattern; it does not respond to intensity as such (73).

In other words, ESP radiation may be following an inverse square law but the subject's response, which depends on interpreting the different parts of the signal, may not necessarily fall off with distance as interpretation involves assessing relative intensities within the overall radiation.

Finally, the evidence that psi is not attenuated by distance can be questioned. As Dobbs puts it:

> First, accepting that telepathy has occurred over great distances, as well as over small distances, it does not follow that distance is wholly irrelevant to its occurrence, or *to the probability of its happening.* We have, for instance, no systematically compiled data to test whether it has happened as frequently over long distances as over short distances, taking into account the number of occasions when it has been tried experimentally (74).

The relative ease with which these arguers overcome the inverse-square-law objection illustrates that even when forced to argue within the confines of specific parts of physics, arguments may still be found to overcome such objections. Indeed, it seems that the closer arguers get to contradictions with hard and fast physical principles the easier they are to avoid by simply making use of the rich variety of ideas already available within physics. Again, no claim for the necessary anti-scientificity of parapsychology can be warranted by reference to the inverse square law.

5. Precognition as a Special Case

If particular physical principles cannot be used to establish an imutable and necessary contradiction with psi-phenomena in general, perhaps it can at least be said that *particular* psi-phenomena do entail such a contradiction. One particular phenomenon of parapsychology — precognition — has been claimed by many arguers to be in necessary contradiction with science. For instance, even if arguers were to feel themselves able to accept the physical possibility of telepathy and psychokinesis, many would still baulk at the possibility of precognition. For example, Knight writes:

'Straight' telepathy poses vast theoretical problems but precognition — with its apparent implication that causation can work backwards in time — seems to violate one of the essential presuppositions of science (75).

Similarly H.H. Price, Ducasse and Rhine himself feel that precognition is "outstandingly incompatible" with physical science:

Precognition seems to require a mode of causation in which the effect occurs *earlier* than the cause, and there is clearly no room for such a process in a Materialistic universe (76).

Of the several modes of extra sensory perception, precognition is the one *prima facie* most paradoxical; for how can a non-existent event — an event that has not yet occurred — cause anything? That an effect, for instance a precognitive dream, should occur earlier than what causes it is a contradiction *ex vi terminorum* and is, therefore, impossible (77).

Nothing in all the history of human thought — heliocentrism, evolution, relativity — has been more truly revolutionary or radically contradictory to contemporary thought than the the results of the investigation of precognitive psi (78).

One way out of this impasse is to re-interpret precognition in terms of some other psi-modality (typically PK — the subject causes his predictions to come about rather than foresees them — a tactic which we have already seen McConnell employ (p. 233). This maintains the arrow of causality without any problem. Indeed, objections to any particular type of psi functioning could be escaped by reinterpreting in terms of another. As Mundle points out:

There has been, and still is, much controversy between psychical researchers as to how many different primary hypotheses must be invoked in order to account for the facts. (79)

Though, as he himself notes, he would not find such a re-interpretation process plausible in all cases; particularly cases where large scale disasters have been precognised, implying that the visionary *caused* the disaster:

It does not seem very plausible to me in view of: (1) spontaneous cases in which the events precognised are major catastrophes. I find it very hard to believe that someone's dream could cause a train crash or an earthquake. (80)

Even if we are not prepared to accept the re-interpretation ploy, and are, therefore, stuck with apparent time reversal, there is still no need to conclude that precognition is incompatible with physics. For instance, some authors have claimed that physics already incorporates time reversal (81), theories advocating more than one dimension for time have been postulated (82), and it has also been suggested that precognition involves 'subjective time' and hence is irrelevant to the objective time with which physics is concerned (83).

An alternative response to the objection that it is logically impossible for an effect to precede its cause is provided by Bob Brier (84) in his book, *Precognition and the Philosophy of Science*. Brier's arguments are neatly summarised by Beloff. First, the logical objection to backward causation:

Let A and B be two hypothetical events which we are taking to be causally related, where A precedes B. Now, once A has occurred, B cannot possibly make it not occur; similarly, if A fails to occur, B cannot then make it occur. Hence, since B can have no effect on A, B cannot be the cause of A. (85)

Brier combats this type of objection by arguing that just the same type of objection can be raised against forward causation where we assume A to be a cause of B. Again in Beloff's words:

... either B will occur or it will not occur. If we assume that B will occur, we cannot also suppose that A can prevent it from occurring; similarly if we assume that B will not occur, we cannot suppose that A can make it occur. Hence A cannot be the cause of B, which is absurd. (86)

Brier concludes after his examination of the logical difficulties of a cause succeeding its effect, that:

> ... even if precognition did necessarily involve backward causation, it cannot be ruled out on the grounds that backward causation is a logical impossibility (87).

Thus, logic alone does not seem to provide an unambiguous answer to the possibility of precognition, and does not therefore make even precognition 'anti-rational' or anti-scientific.

To conclude the discussion of physics and parapsychology, it must be remembered that even if it were decided that they are incompatible, it does not follow that parapsychology and science are incompatible (compare Scriven above, p. 231). Pratt, Ehrenburg and Chauvin in these argument fragments, which are of the format-4 type show that some arguers are happy to accept a 'pluralistic' notion of science which can accommodate psi into science even if not necessarily into physics:

> *Physics is not synonymous with science* ... such an assumption begs the question. If there ultimately should prove to be some range of natural (parapsychical) phenomena that are irreducibly beyond the scope of physics, this state of affairs would not of itself be a contradiction of the concepts of that area of knowledge. It is simply a question of whether parapsychology represents an extension of physics or an extension of science beyond the borders of physics, as that branch can properly be defined. (88)

> Coming from modern physics to parapsychology, I, unlike most of my fellow-parapsychologists, do not consider the two fields of experience incompatible. From my point of view, physics embraces the range of occurrences below the biological level, while psi phenomena transcend mere biology. (89)

> In regard to parapsychology, then, we may affirm only that at the present time we do not know how to reconcile modern physics and psi. But we are not entitled to use the word 'contradict'. The existence of psi does not annul [for example] the laws of electrical currents, for on a proper scale and for the facts they regulate, they are true. (90)

Similarly Tart, in arguing for a special science to study altered states of consciousness (ASC — e.g. states in which psi-phenomena may become manifest) and for pluralistic (state-specific) science in general, does not see the clash with conventional psychology as negating the scientific nature of his enterprise:

> The vast majority of phenomena of ASC's have no known physical manifestations: thus

to physicalistic philosophy they are epi-phenomena, not worthy of study. But insofar as science deals with knowledge, it need not restrict itself to physical kinds of knowledge ... the essence of scientific method ... is perfectly compatible with an enlarged study of the important phenomena of ASC's. (91)

6. *Parapsychology Incompatible With Psychology?*

Certain objectors to psi-phenomena have argued that the incompatibilities lie not (or not only) with physics or science in general, but with psychology. For example, Szasz and Willis write:

The subject matter of psychical research, as has already been noted, is poorly defined. It deals with phenomena which appear to be contrary to *physical* laws. The same idea finds expression in the term ESP ... In this instance, the subject is defined as dealing with matters which appear to be contrary to, not physical, but psychological, laws. (92)

... the conclusions advanced by parapsychologists would be utterly incompatible with the cardinal assumptions on which present-day psychology rests .... (93)

Though we quote no arguments which counteract the incompatibilities proposed by Szasz and Willis directly, their defeasability can be guessed at from Mundle's (94) claim that two of Broad's principles are rejected by most biologists and psychologists and thus that the principles cannot be described as part "of the framework within which all our practical activities and our scientific theories are confined".

The polar opposite position to that of Szasz and Willis is perhaps that of Burt's. It appears that he is putting forward a case for the unification of psi and psychology in pluralistic isolation from the rest of science:

Psychical processes and psychical phenomena (form) ... the very crux of psychology as a separate branch of science .... (95)

This can be regarded as another version of the format-4-type argument.

7. *A Specific Psychological Principle*

In the same way that specific physical principles have been claimed to be incompatible with parapsychological phenomena we find that specific psychological principles can also be held to be in contradiction. Such a claim

Is Anti-Science not-Science? 243

is made by G.R. Price:

There is no learning but, instead, a tendency towards complete loss of ability. (96)

but answered by Meehl and Scriven (among others):

Now it would be reasonable to expect, in a series of experiments intended to show that learning does not occur, some *trial-by-trial* differential reinforcement procedure. Mere continuation, with encouragement or condemnation after *runs of many trials* can hardly provide a conclusive proof of the absence of learning in a complex situation. We ourselves know of *no* experiments in which this condition has been met and which show *absence* of learning; certainly one could not claim that this absence was established. (97)

There is then, nothing in psychology that definitely makes parapsychology anti-scientific.

8. Some Tangential Views

Some idea of the degree of complexity of the compatibility question may, by now, have been given. It is made still more complex by writers' suggestions that even if it were not possible to absorb psi-phenomena into science as it is known, this would not make the phenomena important, but rather it would involve a small or non-existent problem. Thus Scriven suggests:

... the relationship between current physical laws and extra-sensory phenomena is that if accepted, the latter would require that the former be viewed as having a slightly more restricted range ... It is only when the laws are extrapolated from the regions in which they have been directly supported by experimental evidence that they could come into conflict with ESP. (98)

and Thouless and Flew believe:

The demonstration of the reality of ESP, of precognition, and of psychokinesis is a demonstration of the presence of a series of anomalies. (99)

... apart from the anomalous set of very weak effects, everything else is just as it was before. Once the correlations are admitted as exceptions to the various general principles against which they offend, there seems no reason why most sciences (scientists) should be upset further. (100)

In a similar way, critic Boring (101) has said that ESP data represent an "empty correlation" and Stevens (102) that "the signal-to-noise ratio for ESP is simply too low to be interesting". Hoagland (103) has suggested that "unexplained cases are simply unexplained. They can never constitute evidence for any hypothesis", and A.J. Ayer has written:

> The only thing that is remarkable about the subject who is credited with extra-sensory perception is that he is consistently rather better at guessing cards than the ordinary run of people have shown themselves to be. The fact that he also does 'better than chance' proves nothing in itself. (104)

These kinds of comments, which mainly fall within Format 5 of the general arguments we have identified, must be set alongside definitions of psi-phenomena which constitute the field by reference to perceived incompatibilities with current physical knowledge (format 6): Thus Pratt and Beloff suggest:

> Psi-phenomena are precisely those psychological events which defy description in terms of any physical theory now available. (105)

> Parapsychology means the scientific study of the 'paranormal', that is, of phenomena which in one or more respects conflict with accepted scientific opinion as to what is physically possible. (106)

To add a final baroque twist, Scriven (who believes physics is co-extensive with science, see above p. 231), cannot countenance a definition of the physical which is incompatible with the proven facts of parapsychology (format 7):

> The principles of physics do not include those of .... psi, but they must not be incompatible with the behaviour of ... successful psi subjects. (107)

We will conclude our examination of the relationship between scientific ideas and parapsychological ideas at this point. As we have seen, arguers have attempted to find principles and laws either belonging to science as a whole or to specific parts of science with which all of parapsychology or parts of parapsychology can be shown to conflict. Counter-arguments have been posed which seem to deny these incompatibilities. The different formats of argument available suggest that arguers are not compelled to reach any one conclusion over the compatibility issue and its consequences for the scientific status of parapsychology.

Conclusion

We have looked at an area of endeavour, parapsychology, which has often been taken to be associated with anti-science. The lengthy struggle by parapsychologists to gain scientific acceptance for their ideas has precipitated a growing literature concerned with the scientific status of parapsychology and its compatibility with scientific knowledge. This body of argumentation is probably more extensive in the case of parapsychology than other modern so-called, 'anti-science' ideas, for the latter have tended to intersect less with science and hence have attracted less attention (108). (Most 'anti-science' ideas are simply ignored by scientific orthodoxy.) The examination of the compatibility of parapsychological ideas with science has, as we have shown, been inconclusive. A priori reasoning alone, we have argued, has not established, and is unlikely to establish, that the ideas informing parapsychology are incompatible with science.

This being the case, a question remains: Why does the a priori argument have such an appeal, and why has it been pursued at such length? The answer seems to lie in the relationship between science and society. Where science is accepted as the prime source of true knowledge about the world, attacks on knowledge claims may be made by suggesting that they are anti-scientific. It is easier to press such attacks through a priori reasoning than through experiment. The efforts of G.R. Price and C.E.M. Hansel exemplify this position as do some more modern counterparts (109).

Thus, so long as the institutions of science remain prestigious we might expect the majority of those who press the case that parapsychology is anti-scientific to be anti-parapsychology. On the other hand, we might expect most parapsychologists to try to borrow the prestige of scientific institutions. Thus it is no surprise that even where it is claimed that parapsychology and science *do* currently conflict, the claim, when made by a parapsychologist, can be a pro-scientific claim. We have quoted J.B. Rhine as follows:

Nothing in all the history of human thought – heliocentrism, evolution, relativity – has been more truly revolutionary or radically contradictory to contemporary thought than the results of the investigation of precognitive psi. (110)

But Rhine, of course, is an arch-exponent of careful statistical scientific method in the subject. Indeed one sociological study has referred to Rhine's work as 'scientistic' implying a slavish respect for the forms of scientific research (111). What is more, as we have shown elsewhere, parapsychologists have expended a great deal of effort in infiltrating the formal institutions of science such as universities, professional associations and journals (112) — hardly the sort of effort that would be expected of those with anti-scientific interests. It seems then, that perception of the subject as scientific or antiscientific depends not on the relationship between the ideas of psi and science, but upon the arguers interests as regards the legitimacy of parapsychological ideas.

Should science become less prestigious and cease to be seen as the prime source of true knowledge, we might expect the situation to be different; we might expect to find the opposite groups claiming that the same relationships between the ideas of psi and science render parapsychology anti-scientific, or otherwise. Perhaps claims that parapsychology is anti-scientific might also be found within contemporary sub-cultures where science is not valued but parapsychology is.

In this paper, we have not attempted to examine the wider scientific and social contexts which generate interests in the application of labels such as science, anti-science, and non-science. Questions like this must be asked and answered within the wider programme of the sociology of scientific knowledge. In another place, we will try to show the way that laboratory practice, that is scientific action, in part defines the relationship between concepts in this field and science in general (113). Both that analysis, and the analysis pursued here, are only small steps in the wider programme. As such, they may seem tedious, but while it can be thought that a priori arguments can establish the scientific value of knowledge claims, the results of research on the social determinants of scientific knowledge will not be as convincing as they might be.

Acknowledgement

Research for this paper was supported by a grant from the Social Science Research Council, ref. HR 3453/2.

Notes and References

1. For similar notions regarding the relationship of ideas and their use see Shapin's excellent work on Phrenology. Steven Shapin, 'Phrenological Knowledge and the Social Structure of Early Nineteenth-Century Edinburgh', *Annals of Science* **32**, (1975) 219-243 and 'The Politics of Observation; Cerebral Anatomy and Social Interests in the Edinburgh Phrenology Debates', to be published in Roy Wallis (Ed.), *On the Margins of Science, Sociological Review Monographs*, 1979.
2. See P.H. Abelson, 'Pseudoscience', *Science* **185** (1974), and 'Science Beyond the Fringe', Editorial, *Nature* **248**, (1974) 541.
3. See, for example, 'Anti-Science and Pseudoscience', *The Humanist*, July/August 1976; 'The Psychics Debunked', *The Humanist*, May/June 1977; Chris Evans, 'The Occult Revival', *New Humanist*, Sept. 1972; Chris Evans, 'The Occult Revival – Four Years Later', *New Humanist*, May/June 1976; and 'Science and the Paranormal', *New Humanist*, November 1975.
4. K.L. Shewmaker and C.W. Berenda, 'Science and the Problem of Psi'. In J.M.O. Wheatley and H.L. Edge, (Eds.), *Philosophical Dimensions of Parapsychology*, Charles C. Thomas, Illinois, 1976, p. 417.
5. C. Evans, 'Geller Effects and Explanations', *New Humanist*, July/August 1976, p. 53.
6. 'Science Beyond the Fringe', see Note 2.
7. P. Kurtz, 'The Aims of the Committee for the Scientific Investigation of Claims of the Paranormal', *The Zetetic* **1**, No. 1, (1976) 6-7.
8. Quoted in K. Frazier, 'Science and the Parascience Cults', *Science News* **109**, (1976) 346.
9. H.M. Collins, and T.J. Pinch, 'The Construction of the Paranormal'. To be published in Roy Wallis, (Ed.), *On the Margins of Science, Sociological Review Monographs*, 1978.
10. M. Sachs, 'Antiscience within Science', *Humanist*, January/February 1978, 52-53.
11. Collins and Pinch *op. cit.*, 1979 (Note 9).
12. G.R. Price, 'Science and the Supernatural', *Science* **122**, (1955) 360.
13. *Ibid.*
14. C.E.M. Hansel, 'Experiments on Telepathy in Children', *The British Journal of Statistical Psychology* **13**, (1960) 176.
15. S. Toulmin, *The Uses of Argument,* Cambridge Univ. Press, Cambridge, 1958.
16. Quoted in C. Burt, 'The Implications of Parapsychology for General Psychology', *The Journal of Parapsychology* **31**, (1967) 3. (We can rely only on Burt's quotation here, as we have been unable to track down the original source in spite of careful perusal of Hansel's writings.)
17. Price, *op. cit.*, 1955, (Note 12) p. 360.
18. Quoted in C. Burt, 'Psychology and Parapsychology'. In J.R. Smythies (Ed.), *Science and E.S.P.*, Routledge and Kegan Paul, London, 1967, p. 62.
19. We will use the terms 'psi' and 'parapsychology' interchangeably.
20. C.D. Broad, 'The Relevance of Psychical Research to Philosophy', *Philosophy* **24**, (1949) 296.
21. C.W.K. Mundle, Contribution to: 'Symposium: Is Psychical Research Relevant to Philosophy?', *Proceedings of the Aristotelian Society, Supplementary Volume*

24, (1950) 221.
22. H.H. Price, 'Psychical Research and Human Personality', *Hibbert Journal* **47**, (1948-9) 113.
23. Shewmaker and Berenda, *op. cit.* 1976, (Note 4), p. 413.
24. M. Scriven, 'The Frontiers of Psychology: Psychoanalysis and Parapsychology'. In R.G. Colodny (Ed.), *Frontiers of Science and Philosophy,* University of Pittsburgh Press, Pittsburgh, 1962, p. 101.
25. C.E.M. Hansel, 'Experiments on telepathy', *New Scientist* **5**, (1959) 457-459.
26. Price, *op. cit.*, 1948-9, (Note 22), p. 107.
27. C.W.K. Mundle, 'ESP Phenomena, Philosophical Implications of', *The Encyclopedia of Philosophy,* Macmillan and Free Press, New York, **3** (1967) 57.
28. C.J. Ducasse, 'The Philosophical Importance of Psychic Phenomena', *Journal of Philosophy* **51**, (1954) 816.
29. M. Scriven, Contribution to: 'Responses to the Forum on Physicality and Psi', *The Journal of Parapsychology* **24**, (1960) 214.
30. R. Chauvin, 'To Reconcile Psi and Physics', *Journal of Parapsychology* **34**, (1970) 216.
31. J.G. Pratt, Contribution to: 'Physicality and Psi, A Symposium and Forum Discussion', *The Journal of Parapsychology* **24**, (1960) 24.
32. C.W.K. Mundle, 'Some Philosophical Perspectives for Parapsychology', *Journal of Parapsychology* **16**, (1952) p. 267.
33. Mundle, *op. cit.*, 1967, (Note 27), p. 57.
34. Mundle, *op. cit.*, 1950, (Note 21), p. 215.
35. J.R. Smythies, (Ed.), *Science and E.S.P.,* Routledge and Kegan Paul, London, 1967.
36. *Ibid.,* p. 204.
37. *Ibid.,* p. 36.
38. *Ibid.,* p. 225.
39. C.D. Broad, Contribution to: 'Physicality and Psi, A Symposium and Forum Discussion', *The Journal of Parapsychology* **24**, (1960) 16.
40. M. Scriven, Contribution to: 'Physicality and Psi, A Symposium and Forum Discussion', *The Journal of Parapsychology* **24**, (1960) 14.
41. Pratt, *op. cit.*, 1960, (Note 31), p. 24.
42. Scriven, *op. cit.*, 1960, (Note 29), p. 214.
43. R.A. McConnell, 'Physical or Non-Physical?', *Journal of Parapsychology* **11**, (1947) 116.
44. Scriven, *op. cit.*, 1960, (Note 40), p. 14, (his emphasis).
45. G. Murphy, 'Parapsychology: New Neighbour or Unwelcome Guest', *Psychology Today,* May 1968, p. 65.
46. C. Honorton, 'Error Some Place', *Journal of Communication* **25**, (1975) 112.
47. C. Burt, Contribution to: 'Physicality and Psi, A Symposium and Forum Discussion', *Journal of Parapsychology* **24**, (1960) p. 29.
48. *Ibid.,* p. 29.
49. G. Murphy, 'Psychology and Psychical Research., *Proceedings of the Society for Psychical Research* **50**, (1953) 42.
50. McConnell, *op. cit.*, 1947, (Note 43), p. 115.
51. Murphy, *op. cit.*, 1953, (Note 49), p. 42.

52. Mundle, *op. cit.*, 1967, (Note 24), p. 56.
53. C. Burt, 'Parapsychology and its Implications', *International Journal of Neuropsychiatry* **2**, (1966) 373.
54. Price, *op. cit.*, 1948-9, (Note 22), p. 107.
55. McConnell, *op. cit.*, 1947, (Note 43), p. 112.
56. Price, *op. cit.*, 1955, (Note 12), p. 360.
57. See Mundle, *op. cit.*, 1950, (Note 21), p. 222 and Mundle, *op. cit.*, 1967, (Note 27), p. 56.
58. Price, *op. cit.*, 1955, (Note 12), p. 360.
59. J. Beloff, 'Parapsychology and its Neighbours', *Journal of Parapsychology* **34**, (1970) 138.
60. Mundle, *op. cit.*, 1967, (Note 27), p. 56.
61. Price, *op. cit.*, 1955, (Note 12), p. 360.
62. Burt, *op. cit.*, 1967, (Note 18), p. 91.
63. H. Margenau, 'E.S.P. in the Framework of Modern Science', *Journal of the American Society for Psychical Research* **60**, (1966) 221.
64. P.E. Meehl, and M. Scriven, 'Compatibility of Science and E.S.P.?', *Science* **123**, (1956) 14.
65. Burt, *op. cit.*, 1966, (Note 53), p. 374.
66. Mundle, *op. cit.*, 1967, (Note 27), p. 55.
67. Margenau, *op. cit.*, 1966, (Note 63), p. 222.
68. A. Dobbs, 'The Feasibility of a Physical Theory of E.S.P.'. In Smythies, *op. cit.*, 1967, (Note 35), p. 229.
69. J.H. Rush, 'Some Consideration as to the Physical Basis of E.S.P.', *Journal of Parapsychology* **7**, (1943) 48.
70. C.W.K. Mundle, 'The Explanation of E.S.P.'. In Smythies, *op. cit.*, 1967, (Note 35), p. 200.
71. Meehl and Scriven, *op. cit.*, 1956, (Note 64), p. 14.
72. Burt, *op. cit.*, 1966, (Note 53), p. 373.
73. Rush, *op. cit.*, 1943, (Note 69), p. 47 (his emphasis).
74. Dobbs, *op. cit.*, 1967, (Note 68), p. 230.
75. M. Knight, 'Theoretical Implications of Telepathy', *Science News* **18**, (1950) 13.
76. Price, *op. cit.*, 1948-9, (Note 22), p. 112.
77. C.J. Ducasse, 'Causality and Parapsychology', *Journal of Parapsychology* **23**, (1959) p. 95.
78. Quoted in Price, *op. cit.*, 1955, (Note 12), p. 361.
79. Mundle, *op. cit.*, 1950, (Note 21), p. 219.
80. Mundle, *op. cit.*, 1952, (Note 32), p. 266.
81. G. Feinberg, 'Precognition — A Memory of Things Future'. In L. Oteri (Ed.), *Quantum Physics and Parapsychology*, Parapsychology Foundation, New York, 1975, pp. 54-76.
82. C.D. Broad, 'The Philosophical Implications of Fore Knowledge', I and II, *Proceedings of the Aristotelian Society, Supplementary Volume* **16**, (1937) 177-209, 229-245.
83. Margenau, *op. cit.*, 1966, (Note 63).
84. B. Brier, *Precognition and the Philosophy of Science*, Humanities Press, New York, 1974.

85. J. Beloff, Letter to the Editor, *Journal of the Society for Psychical Research* **48**, (1975) 154.
86. *Ibid.*, p. 154.
87. Brier, *op. cit.*, 1974, (Note 84), p. 101.
88. Pratt, *op. cit.*, 1960, (Note 31) p. 24 (his emphasis).
89. W. Ehrenberg, Contribution to: 'Responses to the Forum on Physicality and Psi', *The Journal of Parapsychology,* **24** (1960) 216.
90. Chauvin, *op. cit.,* 1970, (Note 30), p. 217.
91. C.T. Tart, 'States of Consciousness and State-Specific Sciences'. In J.M.O. Wheatley and H.L. Edge (Eds.), *Philosophical Dimensions in Parapsychology*, Charles C. Thomas, Illinois, 1976, p. 446.
92. T.S. Szasz, 'A Critical Analysis of the Fundamental Concepts of Psychical Research', *Psychiatric Quarterly* **31**, (1957) p. 97.
93. Quoted in Burt, *op. cit.,* 1967, (Note 16), p. 3.
94. Mundle, *op. cit.,* 1952, (Note 32), p. 260.
95. Burt, *op. cit.,* 1967, (Note 93), p. 16.
96. Price, *op. cit.,* 1955, (Note 12), p. 360.
97. Meehl and Scriven, *op. cit.,* 1956, (Note 64), p. 14 (their emphasis).
98. Scriven, *op. cit.,* 1962, (Note 24), p. 100.
99. Quoted in J.G. Pratt, 'Some Notes for the Future Einstein for Parapsychology', *The Journal of the American Society for Psychical Research* **68**, (1974) 133.
100. A. Flew, *A New Approach to Psychical Research*, Watts, London, 1953, p. 124.
101. E.G. Boring, 'The Present Status of Parapsychology', *American Scientist* **43**, (1955) p. 113.
102. S.S. Stevens, 'The Market for Miracles', *Contemporary Psychology* **12**, (1967) 1.
103. H. Hoagland, Editorial, 'Beings from Outer Space-Corporeal and Spirutal', *Science* **163**, (1969) 625.
104. A.J. Ayer, 'Chance', *Scientific American,* Oct. 1965, p. 51.
105. Pratt, *op. cit.,* 1960, (Note 31), p. 25.
106. J. Beloff, *New Directions in Parapsychology,* Elek Science, London, 1974, p. 1.
107. Scriven, *op. cit.,* 1960, (Note 29), p. 214.
108. Collins and Pinch, *op. cit.,* 1979 (Note 9).
109. For example see Hanlon's attempt to use 'Occam's Razor' discussed in Collins and Pinch, *op. cit.,* 1978, (Note 9).
110. *Op. cit.,* (Note 78).
111. M.D. Gordon, 'The Institutionalisation of Parapsychology', Unpublished University of Manchester, M.Sc. Thesis, 1975.
112. In *op. cit.,* (Note 9), we discuss the strategy of 'metamorphosis' which could bring parapsychology smoothly into science.
113. Final report to Social Science Research Council on project 'Cognitive Dislocation in Science', 1978. To be published.

ORGANIC FARMERS CELEBRATE ORGANIC RESEARCH: A SOCIOLOGY OF POPULAR SCIENCE

SUZANNE PETERS
McGill University

It is going to take a lot more knowledge to develop a sane, stable agriculture than it did to develop our present, conventional system, just as it took a lot longer to develop the science of ecology than it did mathematics or chemistry. (1)

Today's organic enthusiasts see contemporary organic farming research as an innovative and definitive step to a new agriculture. Science, in their minds, is finally coming to the aid of the organic movement. When closely examined, however, the enthusiasts' sense of 'new science' turns out to have rather curious underpinnings. The research they look to is not only less than conclusive, it is somewhat less than new.

Enthusiasts use the notion 'organic research' to refer to a loosely defined set of scientific and quasi-scientific activities. These include fragments of conventional research, results produced by alternative technology centres, innovations by lay enthusiasts, and much that is promised but not yet delivered. While enthusiasts celebrate the 1970s as a hey-day for organic research, their optimism has a less than certain basis in fact. The movement's 'science' is a collection of disparate sources seen as relevant to their mission.

Enthusiasm has also clouded the organic advocates' sense of history. To the extent that research is presently undertaken, it is better understood not as innovation, but as revival. Science was a primary concern of the very first organic farmers and has been a subject of continual debate throughout the history of the movement. At times enthusiasts have rejected science, and at times they have heralded each new result as *the* major breakthrough. Although no single sentiment adequately characterizes what have been very mixed feelings in the movement, enthusiasts have never ignored the idea of organic research.

In the light of these problems, our understanding of the enthusiasts' 'new science' takes on new dimensions. We need to examine not only the ways in which enthusiasts today construct definitions of organic research, but also the ways in which old, as well as new, notions of science shape the movement's research impulses. Both these issues figure in our understanding of the popular science of organic farming.

Disputed Paternity

Two Europeans fathered the North American organic farming movement: Sir Albert Howard, a British agricultural scientist, and Rudolf Steiner, an Austrian mystic and philosopher. Howard and Steiner, strikingly different in temperament, training and experience, were highly unlikely counterparts. Nevertheless they each independently created a farming system that relied on compost and avoided chemicals: Howard called his the Indore Process; Steiner's instructions became the basis of the Biodynamic Method. Perhaps even more surprising, given their differences, Howard and Steiner shared a similar animosity to conventional science and a sympathy for new forms of investigation.

In 1924, toward the end of his life and at the insistence of a group of his disciples, Rudolf Steiner gave an eight-lecture *Agriculture* course in Koberwitz, Austria. With his agricultural instructions, Steiner hoped to reach "deeper essences" lost to modern man. Agricultural reform was for him only a small part of his search for a new 'Spiritual Science' intended to transcend the limits of conventional 'physical' scientific knowledge. Anthroposophy, his new Christian mysticism, tried to help the farmer regain touch with both spiritual and physical dimensions of the farm, with "Nature as a great totality" (2). By ignoring spiritual perception, in Steiner's view, conventional scientists failed to see the "workings of Nature in all her domains" (3). He made his rejection of conventional science unflinchingly explicit:

The merely intellectual life is not sufficient — it can never lead into these depths. We must begin again from such things. After all, the weaving life of Nature is very fine and delicate. We cannot sense it — it eludes our coarse-grained intellectual conceptions. Such is the mistake science has made in recent times. With coarse-grained, wide-meshed intellectual conceptions it tries to apprehend things that are much more finely woven. (4)

As an alternative, Steiner wanted scientists to adopt new religious guidelines. Here, his sense of mission was crucial; he warned of imminent destruction should these guidelines be ignored.

> Mankind has no other choice. Either we must learn once more, in all domains of life — learn from the whole nexus of Nature and the Universe — or else we must see Nature and withal the life of Man himself degenerate and die (5).

Despite the urgency of his mission, however, Steiner's agricultural instructions were, in practice, closely guarded. As much as he wanted reform, Steiner was as yet unwilling to risk the doubt and scepticism of conventional agriculturalists. He asked his disciples "to exercise the necessary restraint" (6) to keep his intuitions temporarily secret. This cautious approach allowed Steiner to be admittedly esoteric in his instructions — he referred to the deep mystical root of his work on every page of the *Agriculture* course. First he evoked "astral-etheral forces" in the universe, and then he gave instructions which allowed the farmer to use these forces, to prepare "spiritual manures". Given this mysticism, Steiner worried even about the wisdom of exposing his insights to faithful followers.

> I know perfectly well, all this may seem utterly mad. I only ask you to remember how many things have seemed utterly mad, which have none the less been introduced a few years later. (7)

His uneasiness, however, was unfounded. Steiner's instructions aroused no controversy among his followers; their questions, recorded with the course transcript, asked 'How?', 'When?' and 'What?' his instructions required, never 'Why?'. Yet Steiner's disciples listened to his counsel for secrecy. For many years the *Agriculture* course remained under close wraps — individual copies were signed out only for strictly personal use, and no attempts were made to popularize his instructions (8).

At the same time, however, his followers faced a second, seemingly conflicting, mandate: In his hope for eventual public reform, Steiner asked his listeners to try out his insights, "to make experiments in confirmation of these guidelines, and demonstrate how well they can be used in practice" (9). In fulfilling this mission, he told his followers to turn not to agriculturalists, but to peasants. Science needed "peasant wit" or "so-called peasant stupidity". "I have always considered what the peasants and farmers thought about their things far wiser than what the scientists were thinking . . . " (10). The peasant held out the possibility of a new appreciation of the land; Steiner disavowed all attempts to quantify what he saw as essentially empathetic understanding. "For rather would I listen to what is said of his own experiences in a chance conversation, by one who works directly on the soil, than to all the Ahrimanic statistics that issue from our learned scientists" (11). Without the agriculturalist, and with the peasant, Steiner hoped to create a new agricultural science, guarding secrecy, but checking out the usefulness of his intuitions.

After Steiner's death, these responsibilities fell on the shoulders of a small, self-appointed 'research circle' at Dornach, Switzerland, headquarters of Anthroposophy. Almost immediately, small changes began to seep into Steiner's original counsels. An unacknowledged process of secularization and institutionalization took place within the research circle. First, his followers dubbed Steiner's instructions the 'Biodynamic Method', next they numbered his preparations. A spiritual manure, which had been buried over the winter in a cow horn in order to collect astral forces, now became 'Preparation 500'. In this manner the research circle started a slow process of proselytism, avoiding the most visible stigma of mysticism. Similarly, where Steiner had advised peasant wit, and the help of the farmer rather than the scientist, his followers drew on more conventional credentials. Ehrenfried Pfeiffer, a medical student, became the unofficial expert on the Biodynamic Method, trading on his scientific background. Even without Steiner, Anthroposophists still looked for a prophet, if now a more secular one.

Pfeiffer took the Biodynamic Method to the United States. The peasant and farmer, supposedly a new kind of spiritual scientist, nevertheless followed his lead as an acclaimed expert in the research circle (12). In the United States, Pfeiffer received an honorary degree, and became 'Dr Pfeiffer' to the movement. In the 1950s, he developed a commercial compost starter to replace the preparations. His starter was the final secular step away from Steiner's esoteric instructions.

Biodynamicists were not then faithful to Steiner's original vision. To accommodate his dual mandate for secrecy and verification, the movement presented one face to the world, a second to members. Devoted anthroposophists had access to the *Agriculture* course, but outsiders relied only on

Pfeiffer's teachings. Neither group relied on peasant wit.

Sir Albert Howard, our second 'organic father', did not share in Steiner's mysticism. While Howard's two popular appeals for agricultural reform, *An Agricultural Testament* (1940) and *The Soil and Health* (1947), evoke in 'Mother Earth' the Christian sensibilities of his British farming family and in 'the Wheel of Life' the religion of the Far East in which he worked, his spiritual metaphors served a different purpose. Howard was foremost a scientist; his religious appeals were linked to the creation of a popular movement, not to esoteric mysticism. Whatever secular concessions the Biodynamicists made, these were not sufficient for Howard. He always retained his expert posture, denying their claims as "unsubstantiated" and branding them the "muck and mystery school" (13).

Yet Howard was himself disillusioned with traditional research and obsessed with reform. His own disenchantment came from first-hand experience as a British colonial scientist and administrator. In Howard's eyes, most agricultural scientists were "laboratory hermits" (14), who failed to understand the practical needs of the farmer. This conflict characterized his first job in Barbados in 1898. How could he ask the farmer's questions from the laboratory?

> It was borne in on me that there was a wide chasm between science in the laboratory and practice in the field, and I began to suspect that unless this gap could be bridged no real progress could be made . . . (15)

Slowly he began to see the split between his life as a researcher and as a farming advisor as the basic flaw of all agricultural science.

In 1918, now at work in India, Howard began to make plans for a new kind of research station. He hoped to bring science and practice together "without any consideration of the existing organization of agricultural research" (16). His new station began operation at Indore in 1924. Here, Howard perfected his method of composting. His campaign, however, was ideological as well as practical. The station was meant to be a model for a new "synthetic" agricultural science, and his work was intended as a model for "the new investigator" (17).

Howard hoped to become a "brother cultivator" (18) with the peasant. His new investigator would share the practical experience of the farmer, focussing effort on a "single practical problem" (19) on a single piece of land. The scientists would differ from the peasants only in their training and in the wider experience that travel might offer. The first attention of the agriculturalists was to be their farm, and with the peasants their successes were to be 'written on the land'. Like Steiner, quantitative research was anathema to Howard; agriculture was in essence a qualitative concern: "Mother Earth does not keep a passbook" (20).

> Many of the things that matter on the land, such as soil fertility, tilth, soil management, the quality of produce, the bloom and health of livestock, the working relations between master and man, the *esprit de corps* of the farm as a whole, cannot be weighed or measured. Nevertheless their presence is everything; their absence speaks failure. (21)

Howard's new investigator, with his brother the peasant, would develop and cherish an appreciative understanding to his work, care of the land.

Howard found little welcome among his scientific brethren for either the Indore Method or his research reforms. And, as his colleagues rejected his claims, Howard's initial antipathy crystallized. The conventional research establishment became his opposition and he became an enemy within.

After his retirement, Howard turned his efforts away from the scientific community and toward a popular audience. Once only a scientific renegade, he now sought allies in his battle, in the hope that the new investigator might be born of public outcry. The British organic movement, however, paid scant attention to his call for research reform. Although he recruited some auspicious confederates in the medical profession, and the support of members of the landed gentry, most of Howard's support came from gardeners and small-scale farmers. The interests of these enthusiasts were practical, not scientific (22). They were able to romanticize Howard's praise for the peasant in terms of their own daily concerns.

Howard, the scientist, and Steiner, the mystic, broached similar concerns for a new agricultural science. Despite Howard's reservations, much about his notion of the 'new investigator' paralleled the practical emphasis and peasant wit espoused by Steiner. These concerns, however, were largely unimplemented by their immediate disciples. At best, we can say that these reformers shaped not a new science, but a sense of mission for a new agriculture.

An American Stepfather

In the United States, the disparate threads of Howard's and Steiner's reforms were first drawn together by J.I. Rodale, editor and publisher of *Organic Gardening and Farming* magazine. Organic farming was for Rodale a passionate avocation. He was by profession an accountant, owner and operator of a successful electrical wiring company. This commercial success saw his organic crusade through its first troubled years (23).

Rodale began publishing *Organic Gardening* in 1942, full of fierce enthusiasm for research. By 1955, however, his initial optimism was badly tarnished and his efforts to generate organic research were transformed into diatribes against the scientific establishment.

Originally, Rodale took conventional researchers seriously in a way that Steiner or Howard never had. Howard was his hero (24), but Rodale had never experienced any of Howard's research difficulties first hand. He seemed to expect American scientists to be a better, fairer-minded breed than that encountered by his mentor. Somewhat similarly, Rodale's familiarity with Steiner's instructions came once removed, through the secularized and 'expert' teachings of Pfeiffer.

Rodale was influenced by the scientific spirit of American reforms, by the climate of testing generated around the Pure Food and Drug Act, and by the early pesticide controversies. His vision of farming movement depended on a compromise between the researcher and the reformer; ultimately his was a faith in scientific objectivity.

Rodale was a popularizer, not an innovator. *Organic Gardening and Farming* was, and is, a digest, a vehicle for Rodale's research crusade. He compiled in it gems of conventional research, results gathered on his own farm, and the work of other enthusiasts. Straight research was moulded to his private vision, including citations from the United States Department of Agriculture, from university extension bulletins, from agricultural journals, and from classic texts in soil science and agronomy. Rodale also did his own 'research'; following Howard, he treated his farm as an experimental model and made plans for an organic laboratory. Ehrenfried Pfeiffer turned out to be a central collaborator in these efforts; he completed nutritional tests on Rodale's crops, and wrote for *Organic Gardening*. When faced with increasing divisiveness between Howard and the Biodynamicists, Rodale refused to be party to the conflict; he acknowledged both sources in his first book, *Pay Dirt* (1945). His popularizing zeal led Rodale to see all sources as part of one great organic cause.

This same encompassing eclecticism extended to a campaign to recruit establishment scientists to organic questions. In 1949, Rodale began to solicit money from his readers for a Soil and Health Foundation, designed to support researchers sympathetic to a new agriculture (27). There were, however, few applicants for the money that Rodale collected; only scattered bits of research were ever funded. Interested researchers may have been discouraged from taking Soil and Health funds; in at least one rumoured case a scientist accepted a grant only to find out later that the money was contaminated in the eyes of his university administration. The striking exception to this trend was a grant accepted by Drs William Albrecht and Keller at the University of Missouri (28). Keller seems to have been the post-doctoral student enthusiastic about the organic message. Albrecht, more cautious, advocated building humus without condemning all chemicals. William Albrecht, however, was the famous name, and J.I. Rodale made much of his apparent sympathy. Yet the Soil and Health Foundation appeared to have too many strings for most researchers. After 1952, the foundation was a granting agency in name only, and by 1955 it had been unofficially dissolved.

Over the years, Rodale's initial optimism about collaboration with the scientific community began to crumble. "When I began to publish *Organic Gardening* in 1942, I imagined that the agricultural scientists would be waiting for me with open arms, but I certainly did not expect that they would have brickbats in them" (29). As his crusade came under attack by the medical and agricultural establishments, Rodale's equanimity waned. Despite his efforts to create a legitimate scientific basis for the organic farming movement, he found himself branded as a crank and a fraud. The professional opposition, unsympathetic particularly to the nutritional theories of organic advocates, catalogued Rodale alongside its traditional quacks. The brickbats were openly in hand.

The first public confrontation between Rodale and his scientific opponents took place at the 1950 Congressional hearings on Chemicals in Food. Rodale, as the sole contestant for the organic position, got a thorough trouncing. The questions posed turnd his testimony into a quasi-prosecution, a sarcastic cross-examination of his credentials and professional credibility (30). Rodale's evident lack of farming experience, and his obvious dearth of scientific training, became substaintial issues at the hearing.

The conventional scientists seemed particularly incensed by the way that Rodale used published results to support his argument. Discrepancies between his written statement and the stands taken by the researchers he quoted were used against him (31).

Mr Abernathy: Do you know Dr. Emil Truog, of the University of Wisconsin?
Mr Rodale: Yes. In my statement there is a statement by him which says that only through the use of organic matter will you get trace minerals in the soil.
Mr. Abernathy: He testified before the committee and states: Absolutely no authentic evidence exists to the effect that the application of chemical fertilizers to soils in accordance with approved practices, such as are recommended by the various state agricultural experiment stations and the United States Department of Agriculture, causes injury to these soils with respect to their physical, chemical, or biological condition.
Mr. Rodale: I would call that a lie.
Mr. Abernathy: Well, talk to Mr. Truog; not to me.

Thus, although Rodale's inclusive digesting technique swept a wide range of studies under the organic umbrella, he met continual rebuffs from the researchers he embraced.

At the 1950 Hearings, his enthusiastic adaptation of a study by J.K. Wilson of Cornell was directly challenged. Wilson was dead, but the case against Rodale was presented by Richard Bradfield, self-appointed avenger. Bradfield charged that Rodale "was reading into this article ideas of his own which were completely foreign to those of Dr. Wilson." (32). Earlier Bradfield had protested privately to Rodale, but to no avail. Rodale not only continued to refer to Wilson's work, he used the reference in *Organic Gardening* advertisements. Now Bradfield finally got the chance to chastise Rodale publically.

I cite this experience to show the way one prominent leader of the organic school has given his readers an entirely false impression of the work of a competent but unfortunately deceased scientific worker who was not in a position to defend his own views. (33)

The final report of the Committee on Chemicals in Food gave Rodale's 'opinions' only a passing nod. It dealt for the most part with other questions, declaring only that "No reliable evidence was presented that the use of chemical fertilizers has had a harmful or deleterious effect on the health of man or animal" (34).

The Committee was the signal for the scientific establishment to declare open season on the organic idea. The American Medical Association now tuned its position to that of the agriculturalists; its Council on Food and Nutrition solicited a paper from Leonard Maynard, Professor of Nutrition and Biochemistry at Cornell. In his article, Maynard strove to be both thoughtful and judicious; he played a waiting game in which "significant" "final results" must wait "until all the facts are learned" (35). Nevertheless, Maynard's anti-organic sentiments surfaced. He never mentioned Rodale by name, but he hinted at "articles . . . readily recognized as being based on faddism, emotion or propaganda rather than facts" (36). When he called for the "curtailment of enthusiasm based on inadequate data or on speculations" (37) the browbeating taken by Rodale at the hearings came readily to mind.

R.I. Throckmorton, writing on 'The Organic Farming Myth' for *Country Gentleman,* was less cautious about pulling his punches (38). Again Rodale was not named, but Throckmorton did not share Maynard's concern for attaining a tone of scientific reason. He portrayed the organic enthusiast as a gimcrack snake medicine showman, and the movement as a "cult of misguided people" preaching a "strange two-pronged doctrine compounded mainly of pure superstition and myth, with just enough half-truth, pseudoscience and emotion thrown in to make their statements sound plausible to the uninformed" (39). A precis of this article, widely circulated in *The Reader's Digest* in October 1951, under the title 'Organic Farming — Bunk!', downgraded organic claims to an even larger audience (40).

At this point, Rodale began to respond in kind; he had no doubt that he was the intended target of these attacks. Soon his disillusionment with scientists turned into outright animosity. In replying to Throckmorton he blasted the intellectual pretensions of agricultural scientists. "The whole question of the relation of our soils to our health is in the wrong hands" (41). Paradoxically, his faith in science itself was not yet at stake. "This kind of work should be placed with the kinds of brains that created the atomic bomb. Such genius exists, but it is *not* in agricultural science" (42). Rodale still cherished the dream of organic research, but, as his anger grew, he turned against organized science.

In the early 1950s, Rodale began a tentative series of attacks on patronage in agricultural colleges and universities. As early as 1950, he condemned chemical company participation on the boards of governors of agricultural schools. He spoke of his suspicions of a conflict of interest which might ensue when corporations donated research funds to agricultural scientists.

> In my opinion, it would be better if a law were enacted that concerns manufacturing chemical fertilizers and poison sprays not be permitted to donate funds to institutions for agricultural research. That money should come from the Government. Then, I believe, the Universities would not operate under any implied obligations. (43)

In these sporadic attacks, Rodale anticipated something of the environmentalists' political critique of science, but only in a brief and undocumented way.

Despite this newly embittered attitude, Rodale's original faith in science as a helpmate to the organic cause had not entirely dissipated. He renewed his own research efforts at this time; what agricultural scientists would not do Rodale was determined to do for himself. Now some of the money from the Soil and Health Foundation was used to set up the Sir Albert Howard Plots on Rodale's farm. These were a set of 16 cement cylinders, in which both conventional and organic crops were grown. Rodale hoped to compare the nutritional values of crops grown under each method. The results obtained, however, were far from encouraging. When samples were sent out for analysis, the chemically grown vegetables sometimes came out higher in vitamin content. Understandably only the positive results ever appeared in the pages of *Organic Gardening*; the unpublished and damaging figures beautified Rodale's personal files (44). Science was not yet the helpmate Rodale hoped it might be.

In his new efforts, Rodale deviated sharply from some of Howard's central maxims. To Howard, organic and conventional techniques could be compared only on the farm as a whole. Rodale, while retaining the emphasis on the quality of the produce, created plot studies. Howard had specifically condemned these as unsuitable. Rodale was also drawn into the conventional debate about the efficiency of the organic method, arguing for the quantitative as well as qualitative superiority of his techniques.

In any case, Rodale was no longer strictly adhering to Howard's Indore Method. Speaking to American farmers, some on farms of several hundred or thousand acres, he plumbed for ways of placing manures and burned wastes directly on the land. He also opted for rock phosphate, no longer relying, as Howard had done, on the promise that compost-treated soils would break down and make available deeper rock deposits. Here, however, he made only a guarded concession, carefully distinguishing between natural phosphate rock and commercially available phosphate. While the chemical analysis might point up no differences, Rodale suspected unmeasured and undiscerned differences between the two. Yet none of these compromises won Rodale the favour of his conventional critics. In fact, they may have worked against him; unmeasurables were not beloved by his foes, and there were limits to the modifications he would make.

Faced by increasing frustration in this scientific crusade, Rodale took refuge in a new rhetoric of lay understanding. Occasionally, a sympathetic physician or agriculturalist was applauded as one of a rare breed of "maverick scientists" (45), but for the most part the organic farmer was advised to rely on common sense, on his own "unscientific gumption" (46). Just as Howard wanted the scientist to be a farmer, and Steiner preferred peasant to scientific stupidity, J.I. Rodale began to evoke the notion of "common sense" (47) as opposed to scientific evidence.

In this same vein, he now embraced the offensive label crank as a new badge of honour, happy to point out that "cranks turn things around". He congratulated himself as a man of "general intelligence" and urged his readers to do the same. Now, having shared the same ostracism, Rodale's sentiments came in tune with Howard's disdain for the "laboratory hermit".

The changed temper of Rodale's scientific crusade fragmented his following. Many advocates, with his critics, were already doubtful about some of Rodale's fringe interests, including his advocacy of vitamins and health food supplements, his polemics against water fluoridation and aluminium pans. Now enthusiasts were less than unanimous about his agricultural doctrines; some were suspicious of the concessions he made, others of what they saw as his orthodoxy.

From this discontent, Louis Bromfield, a rival agricultural reformer, built an audience for his newsletter, *Friends of the Land* (48). Bromfield, originally a novelist, also published a series of books promoting his vision of a renewed rural America. Again, Rodale confidently reckoned Bromfield as one among many revivalists. Bromfield, however, was weary of Rodale's uncritical enthusiasm. He intended his Malabar Farm as a unique 'scientific' experiment, combining the best of the chemical and organic schools. He disavowed the Biodynamicists, but saw himself as the true inheritor of Howard's mantle. In contrast to Rodale's exile, Bromfield managed to recruit some help and support from the conventional agricultural community (49). He made, however, no serious inroads in their complacency. His impacts were felt largely inside the organic community. Yet most enthusiasts, eager for reform and impatient of Bromfield's slow trial-and-error methods, shared Rodale's doubts about the scientific establishment.

Despite varied attempts at conciliation, the first twenty years of the North American organic movement bore little scientific fruit. Rodale's crusade to recruit conventional scientists, his own 'research', and his endorsement of enthusiasts' claims, all miscarried. Likewise Bromfield's self-conscious compromises were ultimately barren. The enthusiast's popular zeal and his hopes for science remained at odds.

Ecology: The Fairy Godmother

In the early 1960s, environmental concerns seemed to promise a new scientific respectability to the organic movement. For the enthusiast, however, this credibility was born of heavy labour. Robert Rodale, now editor of *Organic Gardening and Farming*, had inherited much of his father's animosity towards scientists, and his faith in common sense. Yet as new sympathies emerged within the scientific community, Rodale, lower-key in style and temperament than his father, began to rethink the problem of organic science. Ecology, and the work of Rachel Carson, became his new science for the movement.

As early as 1960, Robert Rodale was trying out a variety of labels to handle his, and his father's, ambivalence about scientists. In his eyes, it was the scientists, and not the organic farmers, who were cultists. "Their cult is the belief in the superiority of man over nature" (50). In a second editorial that same year, he took a look at the difference between applied and basic research. And, after years of praise for practical research, he came up with an astonishing new conclusion: "that the organic method is a product of basic research primarily — and we seldom quarrel with those scientists who are engaged in basic research" (51).

Yet the Rodales were still suspicious of scientific "tricks". In his 1961 play *The Streets of Confusion*, J.I. Rodale parodied an entomologist, 'Professor

Socially Dexterous', who was "based on some of the scientists I had met at universities and scientific meetings" (52). His entomologist fails to identify a cricket:

> My friend, it isn't as easy as you think. When one is a scientist one has to be scientific, which means one must first hypothesize, then investigate and elucidate and finally correlate, after which one scoffs at, acts superior and supresses. Then and only then can one correctly say that such an insect is such and such an insect. (53)

Robert Rodale took a similar poke at scientific double-talk, paraphrasing what he saw as "the Scientist's 5th Amendent": "I refuse to answer on the grounds that I know what's good for you ordinary people and you don't" (54).

Yet whatever their reservations, the Rodales found themselves in an increasingly anomalous position; not all scientists were so easily burlesqued, and some were voicing their own uneasiness at the implacability of their agricultural colleagues. In 1962, when Rachel Carson published *Silent Spring*, Robert Rodale applauded her as a heroine of the organic movement (55). His father, somewhat begrudging Carson's success, was more sparing of his praise (56). Nonetheless the new environmental crisis foreshadowed by Carson promised much for the organic enthusiast. Ecology, a science "respecting Nature's Laws", held out new hope for the possibility of organic research; Rachel Carson personified the fairy godmother of his dreams.

Carson, however, felt herself a reluctant matriarch. Writing before the bloom of the environmental movement, she was thoroughly uneasy with the organic enthusiast's attempts to link her concerns with his own. She refused to speak to organic groups, and once cancelled a booking on a panel with J. I. Rodale, which had apparently been arranged without her approval (57). Carson felt her position to be tenuous. If anything, she avoided public statements that might portray *Silent Spring* as a diatribe against the chemical industry. Whether or not she covertly sympathized, in no way would Carson jeopardize her credibility by espousing the organic mission. She stated clearly that *Silent Spring* was an attack on the indiscriminate use of pesticides, not on all farm chemicals. Nonetheless her friends and supporters included anthroposophists, and her style and passion echoed Howard's appeal for 'Mother Earth'.

In the years following *Silent Spring*, the mood changed dramatically both inside and outside of the organic camp. J.I. Rodale's initial jealousy of Carson's success moderated, and Robert Rodale's enthusiasm for her efforts increased. The transformation among scientists was even more striking. Carson, originally attacked, was vindicated by a report of the President's Science Advisory Council in 1964. The pesticide controversy, for years the exclusive preserve of entomologists, now came under new public and professional scrutiny.

The organic movement shared in the bounty of Carson's success. As new ecological sympathies emerged, the enthusiast discovered more and more studies which supported his condemnation of conventional agriculture, and more and more allies in his crusade. Environmental activists, politicizing Carson's concerns, precipitated a change in scientific self-consciousness; relevant science became a rallying cry for both the environmental and organic movements. The notions of science "in the public interest" (58) and "for the people" (59) emerged with this new sense of mission. This change in self-consciousness, political and public, as well as personal, revived research as a central issue for the organic enthusiast.

The Exalted Offspring

The 'new science' of the organic enthusiasts springs from both longstanding movement concerns and new environmental activism. Farming questions were new to many among this generation of scientists; agriculture gave them a practical focus for their new sense of purpose, and the organic movement provided a ready made alternative to the problems of chemical farming. In their turn, enthusiasts welcomed these reserarchers warmly; their hopes for organic research now seemed close to fulfillment.

After years of challenge and ridicule, of attack and counter-attack, scientists for the first time came to Rodale Press. John Todd and William McLarney, founders of the New Alchemy Institute, were among the first. It was 1970 when Todd first wrote to Robert Rodale, but this letter still stands out in Rodale's mind: "I get lots of letters, but this was exceptional — it was articulate and informed. So I went up to see what they were doing" (60).

The New Alchemists asked Rodale for money, and in turn they offered a reader research programme. This project treated "The Organic Gardener and Farmer as Researcher" (61); *Organic Gardening and Farming* suscribers were asked to conduct systematic trials which the New Alchemists would collate. The new scientists, however, confronted unexpected difficulties in working with their lay counterparts; many readers made individual modifications to the master plan. It was one thing to plan, another to carry out research by enthusiasts. Robert Rodale recalls the difficulty as one of attainability and applicability of Todd's ideas to the 'average' farmer. 'We want something in which you can tell a farmer building a fishpond how to best dig a 9' by 12' hole. That is not what they are interested in" (62). Direct cooperation between Rodale Press and The New Alchemy Institute was short-lived. Today, neither side wants to go into the story, but their differences arose over both finances and publicity.

In any case, the New Alchemy concept was not only flashier, but more extensive than that envisioned by Rodale. Both were interested in practical solutions, but their sense of the practical was very different. Organic questions were only part of Todd's vision of a new science; Rodale's concerns were both less grand and less encompassing.

Nevertheless the new researchers and the old enthusiasts shared much the same sense of mission. A wide assortment of new research centres emerged, including a new experimental farm at Rodale Press, farm projects at Goddard and Antioch Colleges and at the University of California Santa Cruz, and newer, more modest, versions of Bromfield's Malabar Farm. The researchers at these sites used appreciative as well as technical language. Like Howard, they embraced the farmer as their "brother cultivator" and the farm as their research centre. Thinking of themselves as "stewards", rather than "masters" of the earth, these researchers spoke of "husbandry", not "agriculture", "to remind us of the place of man within the ecological cycle" (63).

Yet many among these scientists also held a new, politicized vision of research. Looking to China, they found a notion of the lay agricultural researcher akin to that of "the barefoot doctor". Chinese research, or "Science that walks on two legs", promised them a new kind of lay researcher; the peasant with a sophisticated knowledge of plant breeding became their model of the new scientist (64). In their own concrete work this political vision created new forms of research organization. Participant research, in their eyes, would be non-hierarchical, based on enthusiasm not credentials, funded inside the movement, and published in popular not technical magazines. In many of the new research centres, this kind of self-consciousness became part and parcel of the call for relevant science.

Despite these inspirations, renewed and new, some things about the movement's research impulse remain much the same. Conventional research results certainly still figure in the enthusiast's new science; in fact they share a great deal with their conventional neighbours. Nick Veeder, a New York teacher who farms organically, raises just this issue: "We need to understand that about 95% of our technology is common, only about 5% is different" (65). Today's organic research still includes much that comes from that shared 95%; many enthusiasts continue to rely on traditional texts for their basic understanding of soils, crops, and livestock, and many of their 'new' results come from the conventional agricultural literature. The rhetoric of the "new science", however, transforms the work of straight scientists. Taking organic and conventional farming systems to be strikingly different entities, the new scientists (researchers as well as farmers) discover hidden significance in conventional results.

This task of melding straight research to the enthusiasts' needs falls to key publicists in the movement. Now, as before, the Rodale Digest performs a central role in this mission, joined today by its new cousins, *Rain, Tilth,* and *Mother Earth News* (66). Recently, however, traditional scientists have become publicists as well.

These new organic 'researchers' lead double lives, splitting their academic and advocate roles. Stuart Hill, of MacDonald College near Montreal, straddles this fence. Hill, a soil ecologist, encourages students to take up organic questions; he hopes for a Centre for Ecological Agriculture at MacDonald which could become the blueprint for a new science (67). As a popularizer, Hill draws from the conventional literature to document organic claims. He travels to organic and conventional conferences with this message, throughout the United States and Canada and recently to the meetings of the International Federation of Organic Agricultural Movements in Switzerland. Other publicists, including Maria Linder, a cancer researcher at Caltech, follow much this same route. Linder's evidence is not her own, but that compiled from Ehrenfried Pfeiffer's early studies (68). Linder's commitment stems from her family life as an anthroposophist, and her summer college work as Pfeiffer's laboratory assistant. Her public involvement, however, has made Linder an invaluable scientific ally in the eyes of the movement. She travels to general conferences and to the meetings of Biodynamicists and organic gardeners. Publicity by these scientific advocates is a large part of what the enthusiast thinks of as organic research.

Yet not all of the researchers embraced by the movement share the enthusiasm of these publicists. Some of this research is not 'organic' at all — it is called 'biological' or 'ecological' agriculture to free it of the taint of attacks directed at the movement. Some of it is only grudgingly organic:

> Originally we were not going to refer to organic and conventional, we were going to refer to experimental and control groups. Then I said, "Look, someone's going to read this, and they are not going to know what we are talking about for a while, but as they read, suddenly a lightbulb is going to go on — and they're going to say, 'These guys are talking about organic farms. Why don't they say so?'". So I decided, let's just say so... It's not a good name, but it's the name it's known by. It's not my fault that the proponents of that kind of farming chose that name. We have to live with that name. (69)

Just as Rodale's digesting technique met with rebuffs, researchers today may or may not revel in their acclaim as organic researchers. Despite growing sympathy for organic farming on the part of many scientists, others are still discomforted by portrayals of their work in the movement. J.D. McLaren's work, at the University of California, Berkeley, is often cited by enthusiasts as proof that whole molecules can be taken up by plants. McLaren, however, is less than convinced that his results warrant this interpretation: "This happened under laboratory conditions, which is not to say that it would happen in the field" (70).

The allegiance of other researchers is hazy at best; some voice concerns about the charge of 'mysticism' leveled at the movement and by implication, at them. William Lockeretz, at the Center for the Biology of Natural Systems in St. Louis, raises this problem: "They are trying to imply that we are somehow suscribers to this mysticism, to the value of organic versus inorganic substances" (71). These fears probably have some basis in fact; the organic movement is rife with horror stories of faithful but persecuted researchers — ridiculed, unpublished, unfunded, unpromoted, and ultimately unemployed. Thus, while the enthusiasts may salute organic farming as "practical magic", some researchers lean to what they see as safer ground.

Similarly, other researchers outline the problem of enthusiastic errors. Bill and Helga Olkowski, scientific heroes to the movement for their activities at the Farallones farm project and the Urban Integral House in Berkeley, do not see themselves as organic advocates.

I wrote him back and I cautioned him to be cautious about the organic farming – how their hearts are in the right place, but their technology needs work. Now I will give you an example: garlic for insecticide. It's an effective insecticide, okay? But if you used it like they used DDT all these years, you'd have the same problems from garlic as will from DDT. Let's say it's like whether you get – whether you have a sledge hammer or a hammer, you still have got the wrong tool in your hand. (72)

The Olkowskis, working on the technology of biological insect control, hope to integrate their work with that of other scientists and enthusiasts without whole-heartedly espousing the organic crusade and its sometimes over-zealous and over-simple solutions.

For the most part, enthusiasts take little note of the private doubts of these reluctant researchers, and few scientists publicly express their reservations. New public dividing lines, however, are slowly being drawn. On the one hand, researchers more carefully distinguish their efforts from those of the enthusiast; on the other, advocates more selectively endorse relevant research.

Energy studies, in particular, have helped to breed new controversies both inside and outside the movement. After 1973, with the public declaration of the 'energy crisis', the organic farming movement and organic research gained new momentum. Comparisons of conventional and organic farming systems now joined the list of new scientific activities extolled by advocates. Yet energy research, initially welcomed, presents problems to enthusiasts.

The first and most widely publicized energy study, carried out by the Center for the Biology of Natural Systems (CBNS) at St. Louis, met a controversial reception. Originally, this study was designed as a general look at agricultural energy consumption. Efforts inside the organic movement were instrumental in changing the research focus; Roger Blobaum, an enthusiast and self-employed agricultural consultant, proposed the organic-conventional comparison to CBNS (73). Blobaum hoped both for vindication of organic methods and a new popular argument for the cause.

At first glance, the study's main conclusions appear ambiguous. For the first year examined, 1973, the CBNs report showed organic farmers at an 8% disadvantage in net profits, for the second year, at a 12% disadvantage. The study suggested, however, that organic farmers used less energy than their conventional neighbours.

Reports of these differences alarmed the agricultural community. The reaction, though less widespread, was as strident as that aroused by Carson's

Silent Spring twelve years earlier. Reviews by agricultural scientists were scathing; Samuel Aldrich of the University of Illinois particularly took it upon himself to provide the basis of the scientific attack (74). In a further step, the National Fertilizer Institute, the industry's Washington lobby, attempted to sabotage the budget granted by the National Science Foundation for the study (75). Nevertheless, the study and its funding were left unscathed.

Yet, ironically, the CBNS study met only mixed reviews inside the movement. These researchers found themselves out of favour with enthusiasts as well as with organic opponents. Intially the reception was warm — enthusiasts discounted the minimal disadvantages shown, and held the hope that proceeding years would narrow the margin. In moving closer to their new science, however, organic advocates began to disdain energy and economic 'proofs' of organic claims (76). The effort to qualify what they saw as essentially qualitative differences left them less than satisfied. The CNBS study, avowed by its investigators to be "standard agricultural economics", failed to get at other differences enthusiasts saw as critical. Even energy or financial advantages would have been inadequate statements of the organic issue in their eyes (77).

Comparative studies undertaken inside the movement point up the problems enthusiasts perceive in economic studies. In 1972, the Rodale Press Soil and Health Foundation was reactivated, and began to fund new projects. Three of these have been non-economic comparative studies, including monoculture versus inter-cropping, the ecological impacts of fertilizers, and dietary differences (78).

In Switzerland, Hardy Vogtmann, director of the Obervil Farm and secretariat of the International Federation of Organic Agricultural Movements (IFOAM), is attempting a comparative study which may soon be copied in North America. Vogtmann is looking at nutritional differences among conventional, Biodynamic, and sheet-composted crops (79). He not only makes the health concerns of enthusiasts explicit, he attempts to win other researchers over to their perspective. In his capacity as IFOAM secretariat, Vogtmann has travelled across the United States and Canada to proselytize for cooperative replicating studies. So far no similar work has been undertaken in North America, but some advocates have expressed interest and are also hopeful thaat the Center for the Biology of Natural Systems may parallel Vogtmann's work. The possibility of this kind of colloboaration, in the hope of new vindicating results, is the latest component of 'organic research'.

Enthusiasts' sense of 'new science' is built on just this kind of optimism. They hope for new technology centres, new politics, new recruits, and new results to bear out their claims. Nevertheless, this 'new science' is characterized by several conflicts. Conventional researchers protest the application of the organic label to their work. Even farmers and researchers who share much the same sense of mission differ about the scope and practicality of research projects. Energy studies in their turn have proved, at best, a mixed blessing, raising as many questions as they resolve for enthusiasts. Some of these differences are easily accommodated, but disputes are intensified as enthusiasts move from practical to comparative concerns.

This discord suggests that organic advocates use the notion 'science' in a very special way. No seminal papers inspire agreement among enthusiasts. Even in the face of external opposition organic advocates seldom unite on the nature of their research claims. Conflicts rather than cooperation characterize their commitment to a 'new science'.

What the organic enthusiasts celebrate as 'new science' is, in fact, their own tenuous popular construction. Few of the 'results' are unique to the organic mission, and those that appear so are often emulations of conventional results. Much of what is celebrated as 'new science' is simply practical technology, conventional or from alternative technology centres, brought to bear on an organic farm.

The birth of what advocates call 'new science' is hailed then, at best, prematurely. When closely examined, we find that enthusiasts have simply adopted old interests to new rhetoric and new purposes. Nevertheless, we cannot deny that the notion of 'new science' serves something very like the role of that elusive progeny for the enthusiast. Organic advocates, like their conventional counterparts, see agriculture as a scientific concern. Their 'new science', whatever its limitations, gives them the chance to recruit new sympathizers, publicize on presentable credentials, and reaffirm their mission. These celebrations of popular science may be more central to the success of the movement than any conventional scholarship might be.

Conclusion

Our exploration of what organic enthusiasts call 'new science' suggests that research means no single thing in the movement. While the enthusiasts see science as a possible means to public acceptability of their crusade, they have had little success with the conventional scientific community. When conventional research fails to fully bear out their claims, these advocates construct their own lay or popular science. They may then claim unmeasurable or intuitive differences (outside the bounds of present scientific assessment), embrace new forms of practical research, and begin to rethink the merits of conventional results. This melange is then called 'new science'.

The birth of this 'new science' goes back to largely unacknowledged traditions in the movement. Our discussion of early advocates suggests that enthusiasts at each stage of the movement have rediscovered science. The two European founders, Howard and Steiner, made research reform a central commitment of their crusades. Both hoped for some new kind of researcher, peasant or scientist, who would be a practical innovator. These early counsels, however, did not resolve the question of science in the organic movement. Almost irresistibly each new generation of enthusiasts has been drawn towards questions reminiscent of those confronted by these first advocates. Only after the conventional scientific community damned his efforts was J.I. Rodale dissuaded from his research crusade. Even then, while he called for 'common sense', his hope for ultimate verification remained transparent. New enthusiasts in the 1960s and 1970s once again seem to be repeating this cycle. Although the main body of their 'new science' is practical, they renew, through attempts at comparative studies of organic and conventional techniques, the hope for public vindication.

Our discussion also suggests that the rhetoric of these debates has had little to do with creating a science of organic farming. No seminal papers or research themes mark out a body of work distinctive to the movement or its scientific sympathizers.

Yet we mistake the character of the enthusiasts' popular science if we simply equate it with 'failed science'. The very endurance of scientific appeals in the movement calls for a different kind of understanding. Most obviously, the tenacity of this popular science suggests the authoritative force of conventional science in shaping the sentiments of the organic movement. Despite the resurgence of religious sentiments and the introduction of new political appeals, enthusiasts persist in their pursuit of science. Less explicitly, the endurance of these appeals suggests that, despite disappointments and disagreements, organic advocates continue to find ways to articulate relatively credible versions of the scientific ethos. The range, richness, and variety of their popular science borrows from the language and credentials of the conventional scientific community. In this manner, enthusiasts successfully project a science-like image to each other and, to a lesser degree, to outsiders. Despite then the limitations of their science, the renewed celebrations of the popular science of organic farming seem to promise enthusiasts realization of the dream of a new agriculture.

Notes and References

1. Dennis King, 'Is Science Advanced Enough for Biological Agriculture?' *Farmstead Magazine*, Summer 1977, Vol. 4, No. 3, p.19.
2. Rudolf Steiner, *Agriculture: A Course of Eight Lectures*, BioDynamic Agricultural Association, Rudolf Stener House, London, 1974. p.118.
3. *Ibid.*
4. *Ibid.*, p.52.
5. *Ibid.*, p.39.
6. *Ibid.*, p.150.
7. *Ibid.*, p.95.
8. Evelyn Speiden Gregg, 'The Early Days of BioDynamics in America', *BioDynamics*, Summer 1976, No. 119, p.26.
9. Rudolf Steiner, *Op. cit.*, (Note 2), p.57.
10. *Ibid.*, p.64.
11. *Ibid.*
12. Pfeiffer's book became the chief U.S. Biodynamic text and his writings made up the *Biodynamics* newsletter. In this latter, he sometimes wrote under pseudonyms, hoping to disguise his dominance in the movement.
13. Sir Albert Howard, *An Agricultural Testament*, Oxford University Press, London, 1940, p.14.
14. Sir Albert Howard, *The Soil and Health: A Study of Organic Agriculture,* (originally published in 1947) first paperback edition, Schoken Books, New York, 1974, p.248.
15. *Ibid.*, pp.1–2.
16. Howard, *Op. Cit.,* (NOte 13), p.40.
17. *Ibid.*, p.221.
18. *Ibid.*, p.222.
19. *Ibid.*
20. *Ibid.*, p.109.
21. *Ibid.*, p.196.
22. The striking exception was the work of Lady Eve Balfour at Haughley Farm. Balfour set up three sub-farms on her land: (1) organic, (2) conventional, (3) mixed. However, she never collected any baseline data on her land, crops, or

animals. Although her 'experiment' was continued by the Soil Association for over 25 years, it was abandoned in 1970 in view of these problems and recent deviations in applying her conditions.
23. As Rodale's widow Anna puts it: "What do you think fed this over all those years?" Personal interview with Anna Rodale June 8, 1977 Emmaus, Pennsylvania.
24. J.I. Rodale, 'Why I started *Organic Gardening*', *Organic Gardening and Farming*, May 1967, Vol. 14, No. 5, p31. 'His idea hit me like a ton of bricks!'
25. J.I. Rodale, *Pay Dirt*, Emmaus, Pa., Rodale Press, 1945.
26. *Ibid.*
27. Interview with Robert Rodale, June 7, 1977 Emmaus, Pa.
28. Transcripts of the Delaney hearings on Chemicals in Food, Washington D.C., 1950, p.864.
29. J.I. Rodale, unpublished mss., Emmaus, Pa. (undated).
30. Transcript, Note 28.
31. *Ibid.*
32. Prepared statement of Dr Richard Bradfield, for the House of Representatives Select Committee to Investigate the Use of Chemicals in Food Products, November 29, 1950, p.7.
33. *Ibid.*
34. Report of the House of Representatives Select Committee to Investigate the Use of Chemicals in Food Products, Washington D.C.
35. Leonard Maynard, 'Soils and Health', *J.A.M.A.* July 1, 1950, p.812.
36. *Ibid.*, p.807.
37. *Ibid.*
38. R.I. Throckmorton, 'The Organic Farming Myth', *The Country Gentleman*, Sept. 1951.
39. *Ibid.*, p.21.
40. R.I. Throckmorton, 'Organic Farming-Bunk!', *Readers Digest*, October 1951.
41. J.I. Rodale, unpublished mss, Emmaus, Pa.
42. *Ibid.*
43. J.I. Rodale, 'Grants to Research Institutions', *The Organic Farmer*, January 1950, Vol. 1, No. 6, p.34.
44. The crop analysis referred to here was done by Edwin Harrington, Agricultural Chemist, Carversville, Pa., April 21, 1953.
45. J.I. Rodale, 'Whither Science?', *The Organic Farmer*, March 1953, Vol. 4, No. 8, p.10.
46. J.I. Rodale, unpublished mss., Emmaus, Pa.
47. J.I. Rodale, *op. cit.* Note 45).
48. *Friends of the Land* was both a newsletter and the name of Bromfield's loose association of fellow organic farmers and land reformers.
49. Louis Bromfield, *From My Experience*.
50. Robert Rodale, 'Have we reached the point of no return?' *Organic Gardening and Farming*, June 1960, Vol. 7, No. 6, p.16.
51. Robert Rodale, 'Is it good to be scientific?', *Organic Gardening and Farming*, Oct. 1960, Vol. 7, No. 10, p.18.
52. J.I. Rodale, editorial notes, *Organic Gardening and Farming*, June 1970, Vol. 22, No. 6, p.104.

53. *Ibid.*, p.105.
54. Robert Rodale, *op. cit.*, (Note 50), p.18.
55. Robert Rodale, 'Rachel Carson's Masterpiece', *Organic Gardening and Farming*, September 1962, Vol. 9, No. 9.
56. J.I. Rodale, *op. cit.*, (Note 24), p.33.
57. Frank Graham, *Since Silent Spring*, Fawcett Crest Books, Greenwich, Conn. 1970: Carleton Jackson, *J.I. Rodale − Apostle of Nonconformity*, Pyramid, New York, 1974, p.34.
58. The Center for Science In the Public Interest is a Washington-based.
59. Science For the People is a Boston-based organization for scientists and science teachers taking a radical look at their own and others' work.
60. Interview with Robert Rodale, June 7, 1977, Emmaus, Pa.
61. John Todd, 'The Organic Gardener and Farmer as Researcher', *Organic Gardening and Farming*, November 1971, Vol. 18, No. 11.
62. Interview with Robert Rodale, June 7, 1977, Emmaus, Pa.
63. Interview with Hardy Vogtmann, July 1977, Montreal, Quebec.
64. Interview with Roger Blobaum, June 16, 1976, St. Louis, Missouri.
65. Nick Veeder is one of 27 American organic farmers with whom I travelled to European organic farms and centres in October 1976.
66. Older revivals, including *Country Journal, Countryside, Farmstead,* and *Acres. U.S.A.* have also captured some of the new organic market since the mid-1960s.
67. Interview with Stuart Hill:
Alan B. Stone, 'Stuart Hill Calls it Ecological Agriculture', *Harrowsmith*, 1977, p.60.
68. Interview with Maria Linder, February 26, 1976, Cambridge, Mass.
69. Interview with William Lockeretz, February 23, 1975, Boston, Mass.
70. Interview with J.D. McLaren, May 27 1976, Berkeley, California.
71. Interview with William Lockeretz, February 23, 1976, Boston, Mass.
72. Interview with William Olkowski, May 28, 1976, Berkeley, California.
73. Interview with Roger Blobaum, June 16, 1976, St. Louis, Missouri.
74. Interview with Don Price, April 24, 1976, Ithaca, New York.
75. This story was picked up in several interviews, some on- and some off-the-record. Because of the assurances I gave to some respondents, I have decided to leave all of them unidentified.
76. Charles Walters Jr., editorial, *Acres U.S.A.*
77. S.B. Hill and J.A. Ramsay, 'Limitations of the Energy Approach in Defining Priorities in Agriculture'. In William Lockeretz (Ed.), *Agriculture and Energy,* New York, Academic Press, 1978.
78. 'Foundation Notes. . .', *The Soil and Health Foundation News,* February 1977, Vol. 6, No. 1.
79. Interviews with Hardy Vogtmann, October 14, 1976, Basel, Switzerland; May 1977, Montreal, Quebec.

HYPER-REFLEXIVITY – A NEW DANGER
FOR THE COUNTER-MOVEMENTS

HILARY ROSE
University of Bradford

I don't know if anyone reads straight through an edited collection of papers such as this volume, or whether, as I confess I usually do, they dip in and out, attracted by a modish title or the author or subject one has some familiarity with already. But to read this book from the beginning is to experience a constellation of voices, some writing directly out of their own personal and subjective experience, all commenting, analysing, attempting to apply the tools of sociology to the phenomenon. Indeed, when we developed the idea of a book on counter-movements in the sciences and invited contributions to it, this was what we anticipated. For the issues raised by such a title, and the difficulties of those trapped within the contradictions of science as it is practised in this society, are multifaceted. We recognised from the beginning that this book would inevitably be as much a declaration of the experience as an analysis of its meaning.

An acute problem for the sociology of the sciences, particularly when it addresses itself to contemporary knowledge, thus becomes that its explanations share the vulnerability of the epistemological and theoretical crises of the knowledges it seeks to analyse. At the same moment that we seek to explore 'Counter-movements in the Sciences' each of us has a personal relationship to these crises and these movements. Sociological writing in this area has both a social and personal significance, which does more than merely notch up another article for the greater glory of the curriculum vitae (1). Indeed, the story of a withdrawn paper makes my point. This, by a historian of science, explored the interconnections between the anti-positivist movement within sociology and anti-authoritarianism. To our disappointment, the author wrote saying that he wanted to withdraw the paper. It was a beautifully written paper, a perfect example, both written from within and also about, one strand of the sociological counter-movement. It became my task to persuade the author to reconsider his decision. He refused on what can only be described as moral and political grounds, in that the paper attributed to positivist sociology an inherent conservatism which he now felt to be unjustified. As a sociologist, I agreed with him; as an editor, I regretted it.

Were there but space enough, several of the issues raised in one form or another by the papers in this collection should be further explored. For example, we should attempt a more thorough-going political sociology of the contemporary counter-movements, placing them historically as Peters and Pearson respectively place the organic farmers and the Luddites. We need to show how they have emerged in response to what Ravetz has called the 'industrialisation of science' (2) or what Steven Rose and I have portrayed as the 'incorporation of science' (3); its capture by the capitalist state which has turned science at an accelerating speed over the last decades from a critical into a dominating force. For, as Böhme reminds us, it is precisely the critique of scientific knowledge itself which separates the 'Old Left' from the 'New' on the question of science. For Bernal and that powerful tradition which followed him, what was wrong with science were the uses to which it was put. Even when, in a vigorous and unfortunately lesser known pamphlet called *Marx and Science* (4) he lays into the class structure of science with a polemical energy which would have endeared him to the New Left, for Bernal, science was in an important sense 'all right'. In the opening pages to *The Social Function of Science* (5) he speaks of the growth of anti-science with a profound revulsion, seeing it as an integral part of the forces of unreason which were to hurl themselves upon the world as Nazism.

The difference today is that most of the counter-movements no longer share this overwhelming confidence in science. What peculiarly troubles these movements (and this I would share) is the nature of scientific rationality in this new world of an incorporated science. The underlying question, which, since we finished writing *Science and Society* (6), has been from time to time a source of fierce debate between Steven Rose and myself, for Steven not only writes about science, he is also a practising biologist, is starkly simple. Is it possible in *this* society to practice science?

Science, even that so called 'basic' or 'pure' science which in the minds of its practitioners was carried out because it was 'interesting' and even because it looked as if it might be socially 'helpful', has a paradoxical way of working itself out as a means of intensifying human oppression. In the areas of 'applied' science, it is embarrassingly evident that experts using what are described as the neutral and objective tools of science, can be called upon to sustain each and every position both of the state and of its critics.

Thus at a fairly powerful gut level I share the anti-science impulse which has manifested itself over the last decade. Yet I live in a society in economic and political crisis and work in a steadily de-industrialising Northern city where each week seems to be marked by some intensified growth of racism and fascism. In these conditions, it becomes increasingly important to understand the nature of the different strands within the counter-movements. What is the relationship of the critique of scientific knowledge to the growth of the cult of unreason? Consciously anti-racist and anti-fascist — for the most part — does the new critique of science, in some presently unidentified way, in practice aid the rebirth of these monstrosities?

The term 'counter-movements' — and the term itself is an uneasy compromise carrying with it the flavour of the counter-culture rather than of critical opposition — embraces then a variety of responses to the oppressive character of an incorporated science.

Some, such as those Helga Nowotny terms pseudosciences, and which Pinch and Collins document for ESP, cry out for incorporation in their turn. By observing the Sanskrit of science, a group of untouchables wishes to translate itself into a position within the caste system of science. Like Black Capitalists or Bourgeois Feminists, they do not wish to change the world — merely to join it. In contrast to this, the radical critique of scientific knowledge not only sees itself as having a transformative agenda, it begins with a much greater openess towards different ways of knowing the world. It is significant that where the Urban Russian revolutionaries of 1917 with their certain faith in science swept away folk medicine, the peasant revolutionaries of China and Africa allied themselves with traditional medicine except where its commitment to superstition was unchangeable. Such an openness to new knowledges brings its own problems, more within the context of the capitalist West than within the context of a struggle for revolutionary survival where the test of practice discriminates rapidly against false knowledge. 'Open-mindedness' in the capitalist West comes at times uncomfortably close to the mere legitimation of petit-bourgeois bohemian life styles. More positively, the openness offers new possibilities both of a refusal of the prevailing scientific rationality with its false claims to neutrality and also of a continuous attempt to develop an alternative rationality derived from the radical needs of the oppressed. Many within the new counter-movements thus see themselves trying to create alternative forms of knowledge which they speak of as 'socialist science', 'feminist science' or 'people's science' in an effort to rehearse within this society at least moments of a future science which no longer dominates nature or humanity.

Thus it remains important for sociologists not only to examine the social formations in which these counter-movements develop, and to analyse the ways in which they interpret and act upon the world; it is also important to ask, who joints the movements, who speaks for them, who are their prophets? One thing that immediately stands out is that, amongst scientists who participate in these movements many, certainly the majority who have contributed to this book, have a background in physics. Why? One explanation is offered by Ruth Hubbard (7), herself a biologist.

The old science is dead. Increasing numbers of people are made anxious and alienated by its proudest success. Proud physics died more than thirty years ago at Hiroshima. Chemistry, which got to be only a little proud in the heyday of nineteenth-century theorising, lost its halo with the advent of dynamite, poison gas and an exuberant chemical industry that belches smelly, deadly pollutants. Biology in its proudest moment is leaving people aghast in the face of new technologies of genetic and developmental manipulation.

That is, chemists lost their illusions too long ago to count, biologists too recently. The defections from physics, then, are likely to be of most significance. Physicists themselves put it differently; here, for instance, is an American physicist's explanation of his apostasy:

Many individuals like myself were attracted to physics in the first place by romantic stories about the golden age of discovery when modern physics was developed by Planck, Einstein, Rutherford, Bohr, Heisenberg, Fermi and their contemporaries. However, today the individual creative genius has been replaced by the large research group, as surely as the mom and pop corner grocery has been replaced by A and P and Safeway, or the local blacksmith by General Motors. The single author experimental paper is as extinct as the Dodo and the single author theoretical paper is becoming as rare as the whooping crane. The style of work at Fermilab or SLAC is little different from that in a large corporation and the typical young high energy experimentalists could easily make the transition to an industrial research laboratory with little adjustment or loss of job satisfaction (assuming he had any to begin with).

There is, however, a more fundamental problem than individual job satisfaction. An industrial system ideally suited to producing tens of thousands of identical automobiles or color television sets, could not have produced the theory of relativity or quantum mechanics, or for that matter, Maxwell's equations, Newtonian mechanics, the theory of evaluation, penicillin, King Lear, the Moonlight Sonata or the Mona Lisa. In other words, as anyone who has ever worked on an industrial assembly line could tell you, efficiency is the antithesis of creativity. It is, therefore, not very surprising that Big Science, for all the investment of money and manpower, has produced precious little that is really innovative and new. It would not be much of an exaggeration to say that the three papers published by Einstein in 1905 did more to change our view of the fundamental nature of the universe than all the work done at Fermilab to date. The Catch 22 of the situation is that even those few physics Ph.D's still able to find work will no longer find that work to be enjoyable, interesting, rewarding or even very productive.

The situation that I have described is apparent to anyone in the physics profession. It will eventually become known to the general public, which is footing the bill, despite a massive public relations apparatus geared up to tout each new resonance found in electron-positron scattering as the greatest advance in human civilisation since the discovery of fire. When those in charge refuse to recognise a disaster as such, then each of us must decide for himself when the time has come to abandon the pumps and run for the lifeboats. I have made my decision and would advise anyone else who can to do the same. (8)

These then are the voices of the recruits to the counter-movements, those who refuse both the alienation of the production of an industrialised science and those who regard it as oppressive. Refugees from *this* science, they join with those non-scientific intelligentsia who have been excluded by C. P. Snow's "men with the future in their bones" (9).

From Critique to Autocritique

Nonetheless, whilst we may begin by acknowledging the widespread retreat from the confident belief that progress in science and technology is associated with social progress, we find ourselves rapidly moving and being moved from a critique of other's knowledges to an autocritique of our own knowledge, and on towards an escalating reflexivity. It is this hyper-reflexivity, spoken of as 'The Disembodied Dialectic' (10), 'on the New Adventures of the Dialectic' (11) and 'The Metaphor Goes into Orbit' (12) which I wish to focus on in this last essay, not least to make my own place clear within the counter- , or as I would prefer to say, oppositional movements to the sciences.

As the new reflexivity, initially a welcome aid to the disenchantment of the sociological world, spiralled through the discourses, it consumed not only 'ideology' but 'science' itself. The certainties of the Althusserian (13) distinction are to be obliterated, scientific knowledges are to be sociologised away, dissolved into social determinations and an equality of discourse. Sociological reductionism has entered (re-entered, Dennis Wrong (14) might not unreasonably argue) sociology. Agnes Heller, writing of Adorno's *Negative Dialectics,* discusses the problems for this hyper-reflexivity, which has difficulty in pointing to the grounds on which it stands in order to make its judgments. She asks,

> What gives the philosopher the power to determine the moment when theory was to be realised or, conversely, the moment when theory failed to be realised? The prophet receives his authority from God. He is justified in pronouncing his knowledge which is hidden from other human beings in so far as God revealed himself to the prophet alone. But if the defetishization of fetishized consciousness is itself judged to be an expression of fetishism, then we can justifiably ask the question: what God enabled this self-proclaimed judge to stand above other human beings and outside of the world, so that he can pass judgment on the consciousness of others and find them guilty of 'fetishism'. (15)

Whilst Adorno is clearly unabashed by playing the part of God's prophet, less assured social analysts are reduced to the assertion that the linguistically and socially constructed nature of 'reality' is a defence against the *"irreality* of chaos and nothingness" which allegedly lies behind all human creations of order (16). For them, in a more modest vein, there is in this irreality nothing to distinguish true from false theories; a new equality prevails between knowledges. 'Authenticity', far from offering to humanity the possibility of engaging in a struggle with problems and thus to defetishise thought and know reality, becomes instead "consciously understanding and admitting the essentially arbitrary nature of the behaviour and identity we choose" (17). To be 'cool', to be aware that we are playing in a series of more or less elaborate games, is the new authenticity.

One of the games players *par excellence* of this new authenticity is Paul Feyerabend (18), a philosopher of science with considerable influence on sociological writing about science, published, for some unfathomable reason by New Left Books. An uninitiated reading of *Against Method* and *Science in a Free Society* would suggest that he is out to destroy scientific rationality, to destroy not only the claims of expertise but Reason and Rationality themselves. In this envisaged 'free society', science, indeed any intellectual activity, simply becomes one tradition among many traditions, one 'ideology' among competing 'ideologies'. "All traditions", he writes, "have equal rights and equal access to education and other positions of power" (19). Not only is Feyerabend distinctly vague about who is to bell the cat of this present unfree society, but his prescriptions for the future society turn on concepts of equality and democracy for all. Why should his 'other traditions', which might wish to advocate inequality and the rule of the strong, share this particular democratic tradition? Why should they be convinced, save on the grounds of precisely the kind of rational social philosophy he invites us to abandon? Not only this, but Feyerabend's method of persuading us to share his views requires precisely that rational logic, albeit interspersed with verbal fireworks, which he professes to oppose. At least Adorno, when faced with the challenge of Popper in debate, refused both to enter the terrain or method (20).

Hence, an initiated reader will see Feyerabend as the self-appointed jester at the court of science. His critique of the formalism of Popperian philosophy is launched from an examination of the practice of science. Thus, it is important to note that the jester criticises the philosopher's conception of scientific method rather than science itself. The practice of science, what Feyerabend usually speaks of as 'research', despite his conviction that all intellectuals except himself should be democratically controlled, remains relatively unscathed. Again, an initiated reader will note the delicacy of his political position. Despite his scientific relativism he separates himself very firmly from that "philosophical relativism" which takes the view that "ideas are equally true or equally false or, in an even more radical formulation, that any distribution of truth values is acceptable". Feyerabend thus manages to act as midwife to this renewed relativism, whilst reserving for himself the social acceptability of the political position of classical liberalism. Whilst *Against Method* was read as the assertion that methodologically speaking 'anything goes', *Science in a Free Society* is to be read as a judicious conflation of radical and conservative social theories. Reminding his readers of his unblemished liberal socialisation from a brief period studying socialist theatre, to his lengthy participation in the special world of Alpbach, the rural Austrian launching-place for the Congress of Cultural Freedom (with all that that signifies about the word 'free'), Feyerabend insists that he is the beloved, if naughty, son of rationality and therefore, of the establishment. His anarchism, he hastens to assure those who believe he has become an enragé, is only the 'medicine' for epistemology. "Anarchism, I say, will heal epistemology and then we may return to a more enlightened and more liberal form of rationality" (21).

But those who do fall into the relativist trap which Feyerabend himself avoids springing, go well beyond Feyerabend's ingenious attack on the scientists who have witch-hunted astrology. Thus a mathematician, Hodgkin (22), writes, "I'd be happy to accept Althusser's definition of scientific practice (working on knowledges to produce new ones) *without* his implication that there is a line which can be drawn separating scientific practices from ideological ones ... (hence) ... astrology done seriously (is a science)". He travels precisely down the path Feyerabend refuses, with only the criterion of 'seriously', which is unspelt out, to ensure that astrology takes its place in the new egalitarianism of knowledges.

However, this is really going too far. Firstly, when Jensen and Eysenck 'work on knowledge' is what they produce new knowledge, and if not, what is it? If it is fetishised consciousness, then we arrive at Adorno's problem, which Heller has already dealt with. If we adopt the position of what is called the 'strong programme' in the sociology of knowledge, then we must presumably regard all these cultural products as new knowledges. Certainly Barry Barnes (23) as an advocate of the strong programme is logically consistent when he calls these new exponents of scientific racism 'new Galileos', for within his framework of sociological relativism, anything does indeed go. If the criterion of truth has been relativised away, the possibility of determining whether this is science or that is ideology has been abandoned, the criterion of 'seriousness' Hodgkin would have us adopt, seems a particularly weak substitute.

Secondly, when I criticise astrology, I am not criticising some future immanent astrology which will spring from the present; I am criticising the present practice of what is called 'astrology', where naive or cynical practitioners consult the equivalent of a ready reckoner to predict whether today is a good day to back a horse, start a love affair or undertake a journey. The 'serious' work on lunar rhythms and corresponding plant or animal responses falls within the problematic of mainstream biology, and does indeed produce new scientific knowledge. But this is not what astrologers do; this work is done by people who describe themselves as biologists. Thus such an abstract defence of astrology is a defence of a knowledge which, while it would be wrong to conclude that it could never be done 'scientifically' or 'seriously' at present shows little or no risk of moving in that direction. Meanwhile, the number of newspapers and especially women's magazines which have a 'stars' page continues to expand. In a secularised society this astrology plays more nearly the part of a religion as 'man's sighs in a heartless world', part of that mystification which denies humanity a chance to realise itself.

The same thesis is spelled out in greater detail in an article by Bob Young entitled "Science *is* Social Relations" (24). I want to discuss this at some length as it represents in a number of ways the paradigmatic case of counter-movement writing within the framework of sociological relativism. It represents the radical wing of the 'strong' programme of the sociology of science and points both to the unresolved difficulties of that theory and also to some rather negative conclusions for those counter-movements it would seek to encourage.

Young's article constitutes a repeated assertion that science is, or may be reduced to, social relations. That is, despite science's claims to be concerned with an understanding of the natural world, it, and indeed any such claims, can only represent a series of social constructs reflective of the social order. In making this statement which, despite its claims to a Marxist lineage, is the antithesis of the traditional Marxist position, Young draws heavily on writings of phenomenology, ethnomethodology and anti-psychiatry. Less well acknowledged, but nonetheless present, is a debt to Feyerabend, and particularly to Sohn-Rethel (25). Sohn-Rethel's thesis is that the emergence of physical science can be linked to the development of abstract thought, itself, he claims, a product of the formation of commodity exchange relationships and the separation of mental and manual labour in, above all, Ancient Greece. Sohn-Rethel thus points to the social origins of science. But he never claims that the existence of social determinants of a phenomenon dissolves the phenomenon itself, and this, as we shall see, is the enterprise which Young sets himself (there are parallels here with the paper by Schmutzer in this volume).

The core of Young's case is his claim that 'the economy and the factory are known by socialists to be social relations'. Hence, by extension, commodities are social relations and, as scientific facts 'are' commodities, they too are social relations. This approach transforms a mediation into an identity and hence allows Young to ignore the multiple forces at play within the factory or the lab. A factory is, at the same moment, part of the actuality of social relations, and is itself objectively real. Its own material reality does not cease because it is part of social reality. Similarly, its products, and the skills of the workers embodied in those products are both real and a part of social relations; they are not coterminous with social relations.

In making this argument, Young does three things. First, he replaces materialism by idealism — only take thought, change our social relationships and the factory will become transformed, the state wither and the millenium arrive. This triumph of the idea is to reverse the achievement of Marx in setting Hegel on his feet — sociological relativism upends Marx and rediscovers a new Hegelianism.

The second mistake derives from Young's understanding of the term 'social relations'. For him, a victim of hyper-reflexivity, the concept is synonymous with *interpersonal* relations — relations between individuals (for instance, of dominance and subservience). Because he defines social relations this way — a very different way from which Marx uses the term, and yet by sleight of hand eliding these different senses — he transforms the slogan of '68 'the personal is the political' into its converse, 'the political is the personal'. But although an apple is a fruit, not all fruits are apples.

His third mistake is to confuse the social determinants of a phenomenon of the phenomenon itself — it is not the social relations of the Hebden Bridge asbestos factory which penetrated the lungs of the workers, but the asbestos fibres. The asbestosis and the painful deaths of the workers are not *merely* social relations either.

To put it more directly, the use of an object is separable from its structure. For example, a house only becomes a house when it is lived in. But the object itself has certain inherent properties that cannot be modified. For example, a house can only be eaten with great difficulty and probably only if alchemy becomes a reality. If objects have inherent properties then it is proper to have an understanding of those inherent properties. That understanding, Marx argues, is a science. Science, of course, involves social relations but the fact that there are social relations does not nullify the fact that if also has a field. To confuse the field with the organisation of the work needed to investigate it is a fatal error that can only lead to the rejection of examining anything that can have the potential of improving life. As Marx argues in his *Critique of the Gotha Programme,* nature has an infinite number of ways in which it can be studied and used, but the way in which we do so is separate from the properties of the particular phenomenon concerned. In other words, what is important is the dialectic that exists between the field and the organisation by which it is used. A failure to make that distinction means that nothing in nature can ever be transformed, and any act of understanding is impossible. The 'authenticity' of critical reason which Heller (26) praised which reaffirmed humanity's capacity, through struggle, to defetishise thought and 'know the thing itself' has clearly ceded to the new authenticity.

The mistake of bourgeois science is to ignore that objects have relationships and histories and are capable of transformation. It is this mechanical materialist reductionism which constitutes the dominant ideology of today's bourgeois science. By contrast, the new idealist ideology pretends that objects do not exist at all, but are merely manifestations of 'social' (i.e. interpersonal) relations, a sort of 20th-century ectoplasm. Within this miasma of sociological reductionism, how can we say that one brick wall is better built than another, let alone one theory or experiment? What it does is to deny both the achievements of human labour, whether these are in bricklaying, cooking or scientific experimentation, and also the autonomy of separate knowledges and the problems of discriminating between and within them. Thus conflict within fields of knowledge reduce solely to 'social' relations, a stance of such monolithic reductionism that paradoxically it enters into complicity with the crudest of economic or biological reductionisms. The political dangers of this sociological relativism — to say nothing of its theoretical inadequacy — are manifest and nowhere more evident than for the counter-movements.

The Onion and the Walnut

Once Marx, invoking the metaphor of a walnut, wrote of revealing the rational kernel within Hegel. For the new sociological relativism the metaphor must be that of the onion. First, reflexivity usefully peels the skin away, then hyper-reflexivity takes over and strips away the remaining layers until nothing — for an onion has no kernel — remains. In keeping with the present preoccupation with personal feelings, this onion-peeling practice is a painful business.

This new subjectivist radicalism stemming from the agonies of intellectuals trapped within an incorporated science tests its theories by their moral fervour rather than by their efficacy. Where Mao (27) was to write of a revolutionist's theory of knowledge:

The Marxist philosophy of dialectical materialism has two outstanding characteristics. One is its class nature: it openly avows that dialectical materialism in the service of the proletariat. The other is its practicality: it emphasises the dependance of theory on practice, emphasises that theory is based on practice and serves practice. The truth of any knowledge or theory is not determined by subjective feelings, but by objective results in social practice. (28)

The new relativists speak of a theory and practice based "in the end" on "personal commitment". Such sociological nihilism, which has gone beyond critique and autocritique, despite all its radical affirmation, reaches out with unseen hands towards an old enemy.

Notes and References

1. Indeed the writing is, to use Alvin Gouldner's title, *For Sociology*, Penguin, 1975.
2. J.R. Ravetz, *Scientific Knowledge and its Social Problems*, Oxford Univ. Press., 1971.
3. Hilary Rose and Steven Rose, 'The Incorporation of Science'. In H. Rose and S. Rose (Eds.), *The Political Economy of Science*, Macmillan, 1976.
4. J. D. Bernal, *Marx and Science*, Marxism Today Series, 9, Lawrence and Wishart, 1952.
5. J. D. Bernal, *The Social Function of Science*, Routledge and Kegan Paul, 1939.
6. Hilary Rose and Steven Rose, *Science and Society*, Allen Lane, 1969.
7. Ruth Hubbard, Introductory chapter to Hilary Rose and Steven Rose (Eds.), *Ideology of/in the Natural Sciences*, Schenkman, (forthcoming).
8. R. J. Yaes, 'The time has come to abandon the pumps and run for the life-boats – Reflections on leaving the Physics Profession to Study Medicine', Talk presented at the Penn. State Conference, August, Am. Inst. Physics, 1978.
9. C. P. Snow, *The Two Cultures*, Cambridge Univ. Press., 1959.
10. D. Carveth, 'The Disembodied Dialectic: A Psychoanalytic Critique of Sociological Relativism', *Theory and Society* 4, 73-101 (1977).
11. A. Heller, 'On the New Adventures of the Dialectic', *Telos*, pp.134-142, 1977.
12. Hilary Rose and Steven Rose, 'The Metaphor Goes into Orbit: Science is *Not* All Social Relations', *Science Bulletin*.
13. L. Althusser, *For Marx*, Allen Lane, 1965.
14. D. Wrong, 'The Over Socialised Conception of Man in Modern Sociology', *Am. Soc. Rev.* 26, 183-193 (1961).

15. A. Heller, *op. cit.*, 1977.
16. I am much indebted to Carveth's paper for this insight into the underlying 'irreality and chaos' embedded within the new sociological, or as he rightly terms it, 'sociologistic reductionism', pp. 10, 197.
17. D. Carveth, *op. cit.*, p. 197.
18. P. Feyerabend, *Against Method,* New Left Books, 1975.
    P. Feyerabend, *Science in a Free Society*, New Left Books, 1978.
19. P. Feyerabend, *op. cit.*, 1978.
20. T. Adorno, *The Positivist Dispute in German Sociology,* 1972.
21. P. Feyerabend, *op. cit.*, 1978.
22. L. Hodgkin, 'A Note on Scientism', *Radical Science Journal* 5, 8 (1977).
23. B. Barnes, *Scientific Knowledge and Sociological Theory*, Routledge and Kegan Paul, 1974.
24. R. Young, 'Science *is* Social Relations', *Radical Science Journal* 5, 65-129 (1977).
25. A. Sohn-Rethel, *Mental and Manual Labour,* Macmillan, 1976.
26. A. Heller, *op. cit.*, 1977.
27. Mao Tse-Tung, 'On Practice', July, 1937. *Selected Works of Mao Tse-Tung,* People's Publishing House, pp. 295-305, 1965.
28. M. Young, 'Taking Sides Against the Probable: Problems of Relativism and Commitment in Teaching and the Sociology of Knowledge', *Educational Review* 25, (3) 221 (1973).

INDEX

Albert, H.   108
alternative technology   32, 251
Althusser, Louis   46-48, 282, 284
anti-science   1-8, 23 n4, 28, 95-96, 107, 122 n5, 221, 223, 245, 279
  romantic tradition of, 13-15, 18
astrology   223, 284

Beloff, J.   235, 240, 244
Bernal, J.   106, 174, 176, 184 n57, 278
Black Act,   197-199
Böhme, Gernot   x, *105-125*, 278
Bradfield, R.   259
British Society for Social Responsibility on Science   29-30, 35
Broad, C. D.   228, 232
Bromfield, Louis   262-263
Bunge, Mario   71, 74-76, 79, 98 n35, 101 n69
Burt, C.   233-236, 238, 242
Bush, Vannevar   69
Bythell, D.   191

Capra, F.   103 n83
Carson, Rachel   264
categorical imperative   59, 65
Center for the Biology of Natural Systems   269-270
chartism   192, 205-206
Committee for the Scientific Investigation of Claims of the Paranormal   5, 222
Committee on Chemicals in Food   258-260
complex problems   151-153, 157
Copernicus, N.   164-166

counter-movements in the sciences,   1, 9, 127-130, 139, 144, 278-281
  social rationality of   2-6, 12-23
  social scientists and   138-140
critical science   3, 21-23, 52, 105-106

demarcation criteria   8, 67, 74, 79, 90-92, 109
  experiments as   74, 77
Dirac, P. A. M.   77, 92
do-it-yourself science   17-18
Ducasse, C. J.   230-232, 239
Duerr, H. P.   96

ecology   263-265
Engels, Friedrich   188-189, 204-205
Epistémé   45, 61, 62

Feinberg, G.   100 n51
Feyerabend, Paul   80, 96, 163, 167, 173, 282-285
Finalisation in Science   116-119, 121
folk science   9
Franck, Robert   viii, *39-56*
frontier science   67-85, 93, 156

Galileo, G.   44, 58, 64, 94, 160
Gardner, Martin   74-76, 82-83
Goethe's theory of colours   110-111
Goldsmith, Teddy   35
Geller, Uri   222
Greenberg, Dan   24 n8
Grabner, Ingo, and Reiter, Wolfgang   ix, *67-104*
Grupp, Michael   ix, *147-160*

291

Hansel, C. E. M. 226, 230
Harper, Peter 32
Harris, Marvin 25 n24
Heller, Agnes ix, *57-66*, 282, 287
Hill, Christopher 186
Hobsbawm, Eric 186, 195
Howard, Albert 252-256
Huxley, Aldous 107
Hume, David 224
Husserl, E. 57, 62, 64
hyper-reflexivity in the sciences 272-289

Illich, Ivan 140

Jordan, P. 82

Kant, I. 59-61, 149, 225
Keegan, G. J. 100 n58
Kurtz, Paul 222

Levy-Leblond, Jean-Marc 17, 20, 81, 93
Luddites 188-189, 195-196, 211, 213, 216 n17

Marcuse, H. 122 n8
Margenau, H. 235-237
Marx, Karl 165, 286-287
Marxism 107, 109, 285
Maynard, Leonard 260
McConnell, R. A. 233-234, 239
McKeown, T. 172
Mumford, L. 172
Mundle, C. K. 229-232, 234-235, 237, 240, 242

new alchemists 265-266
Newton, I. 28, 111, 225
non-science 1
Nowotny, Helga viii, *1-26*

objective knowledge and subjective opinions 40-44, 49-50, 54-55
official science 9, 15
O'Neill, G. K. 101 n61

organic farming movement 252-271
organic farming research 251-271
Organic Gardening and Farming *257*, 263, 265

para-psychology 16, 73, 76, 81, 221-246
  incompatible with physics 231-242
  incompatible with psychology 242-245
  incompatible with science 227-231
particle accelerators 83-85
particle physics 73, 81, 85, 94
  division of labour in, 85-86
Pearson, Geoffrey xi, *185-220*, 278
Peel, Robert 194
Peters, Suzanne xii, *251-273*, 278
Pfeiffer, Erwin 254, 267
Pinch, Trevor, and Collins, Harry xi, *221-250*, 279
Plumb, J. H. 190-192, 215 n9
Popper, Karl 164
Poser, H. 108
positivism 60, 278
Price, G. R. 225, 228, 235, 243
Price, H. H. 229-230, 232, 235, 239
pseudo-science 1, 4, 6, 15-18, 67, 73-75, 80-82, 109, 167, 279
psychoanalysis 73-75, 81, 109, 115

quarks 77-78

Radical Science Journal 31
Radnitzky, G. 108, 122 n11, 123 n16
rationality 4-8, 83, 144, 212,
  of industrial society 127, 140
  of the sciences 130, 132, 138
  scientific 4-5, 95, 136
  social 5-6
Ravetz, Jerome viii, *27-37*, 278
reason/nature dualism 44-48
Report from the Iron Mountain 43
Resurgence 34-36
Rhine, J. B. 224, 239, 245-246
Rodale, J. I. 257-263, 265, 272
Rodale, Robert 263-265

Rose, Hilary   xii, *277-289*
  and Steven Rose,   31
Rose, Steven,   278
Russell, Bertrand   171, 229

Sachs, M.   223
Sahlins, M.   171-172
Saint-Simon, H. de   132
Salpeter, E. E.   88-89
Schmutzer, Manfred E. A.   xi, *161-184*, 285
Schrödinger, Erwin   163-164
Science for People   27, 29-31
Schumacher, E. F.   34
science, alternatives in   105-119
  alternatives to   109-112, 169
  and religion   28, 46-48, 70-72, 153-154, 172,
  demarcation of   7, 68, 74, 109
  hegemony of   2, 9-10, 20, 67, 82-83, 94-95, 174-177
  ideology of   10, 47, 50-52, 84, 108, 131
  interests in   105-106
  myth of   72, 75, 83, 91-92, 95
  social organisation of   106, 156-157
sciences, natural   50, 59, 62, 65, 115, 130
  language of   58-64
  social   61-62
  sociology of   vii, 8, 277
scientisation of the world   127-146
Scriven, Michael   229, 231-233, 236, 243-244

simple problems and the scientific experiment   148-150, 168
Smelser, Neil   189, 193
Sneed, J. B.   117
Sohn-Rethel, Alfred   165-166, 285
Starnberg group,   ix, 105
Steiner, Rudolph   252-254
subject/object dualism   44-48, 55

tachyons   77-78, 82
Taylor, W. Cooke   205-206
The Ecologist   35
Thom, René   111, 167
Thompson, Edward   xi, 186, 190, 196-197, 200, 203-204
Throckmorton, R. I.   260
Todd, John   265-266
Truzzi, Marcello   222

Ullrich, Otto, ix, *127-146*, 159 n9
Undercurrents   27, 31-35

van den Daele, W.   110
value judgements   40-41, 43, 49-50, 55
Velikovsky, I.   72-73, 221
von Weizsäcker, C. F.   148

Weber, Max   127, 175
Weisskopf, V. F.   70-71, 102 n74
Williams, Raymond   186, 208

Young, Robert   31, 285